MORE

Practice Papers for SQA Exams

Standard Grade I Credit

Mathematics

01/071010

ISBN 978-1-84372-874-0

Published by
Leckie & Leckie Ltd,
An imprint of HarperCollins*Publishers*
Westerhill Road, Bishopbriggs, Glasgow, G64 2QT
T: 0844 576 8126 F: 0844 576 8131
leckieandleckie@harpercollins.co.uk www.leckieandleckie.co.uk

Special thanks to
Exemplarr (layout and illustration),
Caleb O'Loan (proofread), Jill Laidlaw (proofread)

A CIP Catalogue record for this book is available from the British Library.

Introduction

More Practice!

This book contains brand new practice exams, which mirror the actual SQA exam as closely as possible in question style, level, layout and paper colour. It is a perfect way to familiarise yourself with what the exam papers you will sit will look like.

The answer section at the back of the book contains fully worked answers to each question, letting you know exactly where marks are gained in an answer and how the right answer is arrived at. It is also packed with explanatory notes, helpful advice and hints to maximise your understanding of the types of questions you're likely to face in the exam. The answers also include helpful cross-references to Leckie & Leckie's book "Credit Maths Revision Notes".

As the name suggests, this book is a volume of *More* Practice Exam Papers for Standard Grade Credit Level Maths. Its sister publication (ISBN: 978-1-84372-773-6 Standard Grade Credit Maths Practice Papers for SQA Exams) is also available and is packed full of entirely different practice exams, worked solutions and helpful explanations, hints and exam tips.

How To Use This Book

The Practice Papers can be used in two main ways:

1. You can complete an entire practice paper as preparation for the final exam. If you would like to use the book in this way, you might want to complete each practice paper under exam-style conditions by setting yourself a time for each paper and answering it as well as possible without using any references or notes. Alternatively, you can answer the practice paper questions as a revision exercise, using your notes to produce a model answer. Your teacher may mark these for you.

2. You can use the Topic Index at the front of this book to find all the questions within the book that deal with a specific topic. This allows you to focus specifically on areas that you particularly want to revise or, if you are mid-way through your course, it lets you practise answering exam-style questions for just those topics that you have studied.

Revision Advice

Work out a revision timetable for each week's work in advance – remember to cover all of your subjects and to leave time for homework and breaks. For example:

Day	6pm–6.45pm	7pm–8pm	8.15pm–9pm	9.15pm–10pm
Monday	Homework	Homework	English Revision	Chemistry Revision
Tuesday	Maths Revision	Physics Revision	Homework	Free
Wednesday	Geography Revision	Modern Studies Revision	English Revision	French Revision
Thursday	Homework	Maths Revision	Chemistry Revision	Free
Friday	Geography Revision	French Revision	Free	Free
Saturday	Free	Free	Free	Free
Sunday	Modern Studies Revision	Maths Revision	Modern Studies Revision	Homework

Make sure that you have at least one evening free a week to relax, socialise and re-charge your batteries. It also gives your brain a chance to process the information that you have been feeding it all week.

Arrange your study time into one-hour or 30-minute sessions, with a break between sessions, e.g. 6pm–7pm, 7.15pm–7.45pm, 8pm–9pm. Try to start studying as early as possible in the evening when your brain is still alert and be aware that the longer you put off starting, the harder it will be to start!

Study a different subject in each session, except for the day before an exam.

Do something different during your breaks between study sessions – have a cup of tea, or listen to some music. Don't let your 15 minutes expand into 20 or 25 minutes though!

Have your class notes and any textbooks available for your revision to hand as well as plenty of blank paper, a pen, etc. You should take note of any topic area that you are having particular difficulty with, as and when the difficulty arises. Revisit that question later having revised that topic area by attempting some further questions from the exercises in your textbook.

Revising for a Maths exam is different from revising for some of your other subjects. Revision is only effective if you are trying to solve problems. You may like to make a list of 'Key Questions' with the dates of your various attempts (successful or not!). These should be questions that you have had real difficulty with.

Key Question	1st Attempt		2nd Attempt		3rd Attempt	
Textbook P56 Q3a	18/2/10	×	21/2/10	✓	28/2/10	✓
Practice Exam D Paper 1 Q5	25/2/10	×	28/2/10	×	3/3/10	
2008 SQA Paper, Paper 2 Q4c	27/2/10	×	2/3/10			

The method for working this list is as follows:

1. Any attempt at a question should be dated.
2. A tick or cross should be entered to mark the success or failure of each attempt.
3. A date for your next attempt at that question should be entered:
   for an unsuccessful attempt – 3 days later
   for a successful attempt – 1 week later
4. After two successful attempts remove that question from the list (you can assume the question has been learnt!)

Using 'The List' method for revising for your Maths Exam ensures that your revision is focused on the difficulties you have had and that you are actively trying to overcome these difficulties.

Finally, forget or ignore all or some of the advice in this section if you are happy with your present way of studying. Everyone revises differently, so find a way that works for you!

Transfer Your Knowledge

As well as using your class notes and textbooks to revise, these practice papers will also be a useful revision tool as they will help you to get used to answering exam-style questions. As you work through the questions you may find an example that you haven't come across before. Don't worry! There may be several reasons for this. The question may be on a topic that you have not yet covered in class. Check with your teacher to find out if this is the case. Or it could be that the wording or the context of the question is unfamiliar. This often happens with reasoning questions in the Maths exam. Once you have familiarised yourself with the worked solutions, in most cases you will find that the question is using mathematical techniques which you are familiar with. In either case you should revisit that question later to check that you can successfully solve it.

Trigger Words

In the practice papers and in the exam itself, a number of 'trigger words' will be used in the questions. These trigger words should help you identify a process or a technique that is expected in your solution to that part of the question. If you familiarise yourself with these trigger words, it will help you to structure your solutions more effectively.

Trigger Words	Meaning/Explanation
Evaluate	Carry out a calculation to give an answer that is a value.
Hence	You must use the result of the previous part of the question to complete your solution. No marks will be given if you use an alternative method that does not use the previous answer.

Trigger Words	**Meaning/Explanation**
Simplify	This means different things in different contexts: Surds: reduce the number under the root sign to the smallest possible by removing square factors. Fractions: one fraction, cancelled down, is expected. Algebraic expressions: get rid of brackets and gather all like terms together.
Give your answer to . . .	This is an instruction for the accuracy of your final answer. These instructions must be followed or you will lose a mark.
Algebraically	The method you use must involve algebra, e.g. you must solve an equation or simplify an algebraic equation. It is usually stated to avoid trial-and-improvement methods or reading answers from your calculator.
Justify your answer	This is a request for you to clearly indicate your reasoning. Will the examiner know how your answer was obtained?
Show all your working	Marks will be allocated for the individual steps in your working. Steps missed out may lose you marks.

In the Exam

Watch your time and pace yourself carefully. You will find some questions harder than others. Try not to get stuck on one question as you may later run out of time. Rather, return to a difficult question later. Remember that if you have spare time towards the end of your exam, use it to check through your solutions. Mistakes are often discovered in this checking process and can be corrected.

Become familiar with the exam instructions. The practice papers in this book have exam instructions at the front of each exam. Also remember that there is a formuae list to consult. You will find this at the front of your exam paper. However, even though these formulae are given to you, it is important that you learn them so that they are familiar to you. If you are continuing with Mathematics next session it will be assumed that these formulae are known in next year's exam!

Read the question thoroughly before you begin to answer it – make sure you know exactly what the question is asking you to do. If the question is in sections, e.g. 15a, 15b, 15c, etc, then it is often the case that answers obtained in the earlier sections will be used in the later sections of that question.

When you have completed your solution read it over again. Is your reasoning clear? Will the examiner understand how you arrived at your answer? If in doubt then fill in more details.

If you change your mind or think that your solution is wrong, don't score it out unless you have another solution to replace it with. Solutions that are not correct can often gain some of the marks available. Do not miss working out. Showing step-by-step working will help you gain maximum marks even if there is a mistake in the working.

Use these resources constructively by reworking questions later that you found difficult or impossible first time round. Remember: success in a Maths exam will only come from actively trying to solve lots of questions and only consulting notes when you are stuck. Reading notes alone is not a good way to revise for your Maths exam. Always be active, always solve problems.

Good luck!

Topic Index

Topic	D Paper 1	D Paper 2	E Paper 1	E Paper 2	F Paper 1	F Paper 2	Knowledge for Prelim			Knowledge for SQA Exam		
							Have difficulty	Still needs work	OK	Have difficulty	Still needs work	OK
Number & Money												
• Decimals	Q2		Q2		Q2							
• Fractions	Q3		Q1		Q1							
• Percentages	Q1			Q3, Q4		Q1						
• Ratios				Q10								
• Scientific Notation		Q3				Q2						
Shape & Measure												
• Bearings		Q8		Q7		Q7						
• Circles	Q9	Q11b	Q13	Q6		Q9						
• Pythagoras' Theorem	Q9	Q4	Q10, Q13									
• Similarity		Q5	Q7		Q6	Q10						
• Solids				Q13a		Q6						
Trigonometry												
• Area Formula		Q2		Q9								
• Cosine Rule						Q7						
• Right-angled Triangles		Q11a				Q4						
• Sine Rule		Q8		Q7								
• Trig Equations		Q10		Q8		Q11						
• Trig Graphs			Q12									
Algebraic Relationships												
• Algebraic Fractions	Q11b		Q3		Q3b	Q10						
• Brackets	Q7b		Q6	Q11	Q7	Q5a, Q8b						
• Factorisation	Q12	Q7b		Q11	Q3a	Q8c						
• Formulae	Q11		Q3	Q13b	Q11	Q8						
• Function Notation				Q12	Q4							
• Indices			Q8, Q14b			Q5b						
• Inequalities				Q5	Q11c							
• Linear Equations	Q5	Q7a			Q7							
• Quadratic Equations	Q12	Q6	Q10	Q1, Q11								
• Sequences/nth terms	Q7		Q14									
• Simultaneous Equations			Q9		Q10							
• Substitution				Q12	Q4							
• Surds	Q10		Q10, Q11			Q5c						
• Variation			Q11									
Graphical Relationships												
• Linear Graphs	Q4	Q9	Q5		Q5							
• Quadratic Graphs				Q12	Q9							
Statistics & Probability												
• Probability	Q6		Q4									
• Standard Deviation & Quartiles	Q8	Q1		Q2		Q3						

Practice Exam D

Mathematics | Standard Grade | Credit

Practice Papers
For SQA Exams

Exam D
Credit Level
Paper 1
Non-calculator

You are allowed 55 minutes to complete this paper.

Do **not** use a calculator.

Try to answer all of the questions in the time allowed, including all of your working.

Full marks will only be awarded where your answer includes any relevant working.

FORMULAE LIST

The roots of $ax^2 + bx + c = 0$ are $x = \dfrac{-b \pm \sqrt{(b^2 - 4ac)}}{2a}$

Sine Rule: $\dfrac{a}{\sin A} = \dfrac{b}{\sin B} = \dfrac{c}{\sin C}$

Cosine Rule: $a^2 = b^2 + c^2 - 2bc\cos A$ or $\cos A = \dfrac{b^2 + c^2 - a^2}{2bc}$

Area of a triangle: $\text{Area} = \dfrac{1}{2}ab\sin C$

Standard Deviation: $s = \sqrt{\dfrac{\Sigma(x - \overline{x})^2}{n-1}} = \sqrt{\dfrac{\Sigma x^2 - (\Sigma x)^2/n}{n-1}}$, where n is the sample size.

DO NOT WRITE IN THIS MARGIN

	KU	RE

1. Evaluate 35% of £351. — KU 2

2. Evaluate $4{\cdot}25 \div (4{\cdot}3 + 0{\cdot}7)$. — KU 2

3. Evaluate $1\frac{1}{2} - \frac{3}{5}$ of $1\frac{2}{3}$. — KU 3

4.

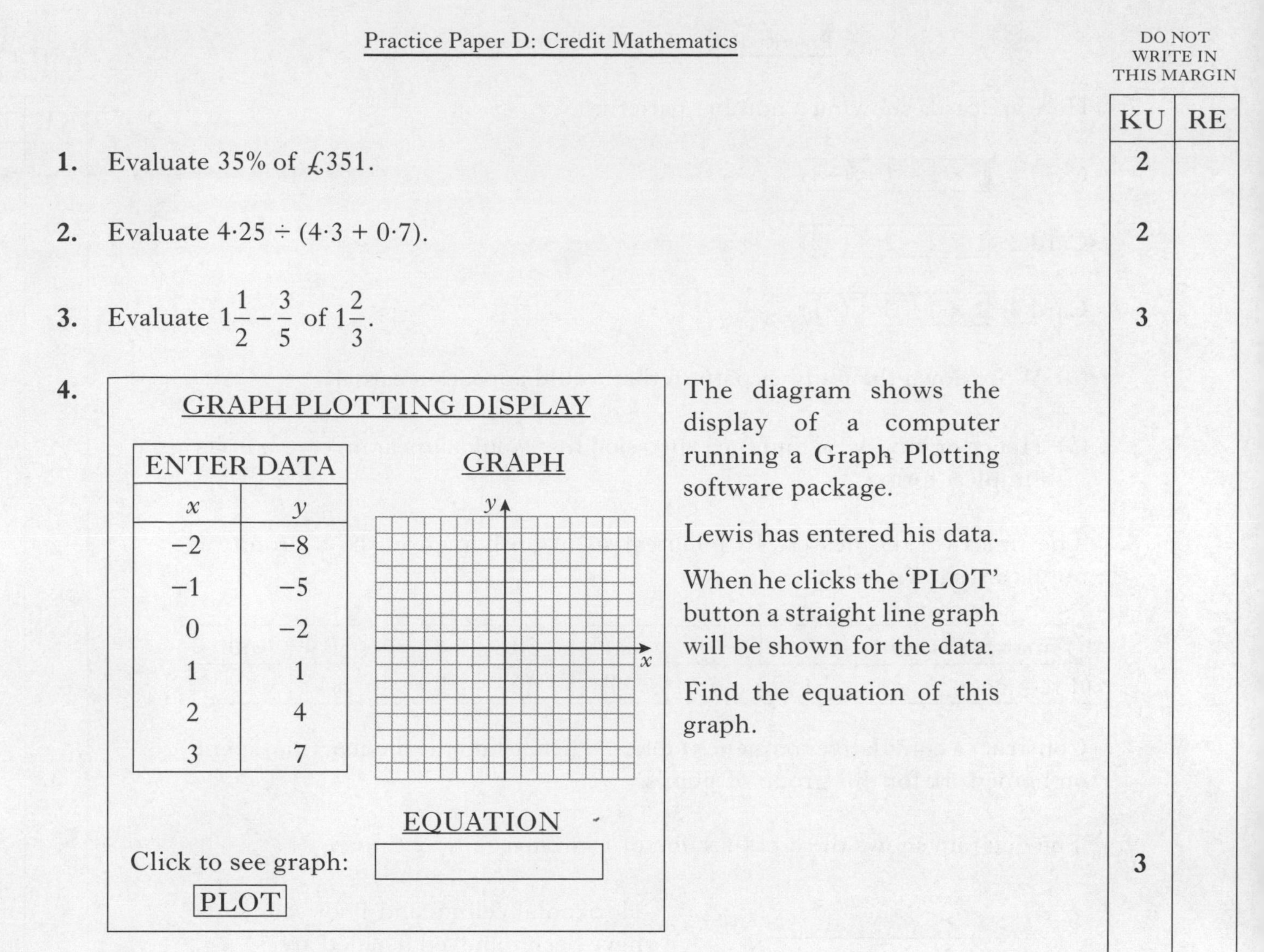

x	y
−2	−8
−1	−5
0	−2
1	1
2	4
3	7

The diagram shows the display of a computer running a Graph Plotting software package.

Lewis has entered his data.

When he clicks the 'PLOT' button a straight line graph will be shown for the data.

Find the equation of this graph. — KU 3

5. Solve the equation $3 + \frac{2}{k} = 1$. — KU 3

6. Two identical spinners are shown:

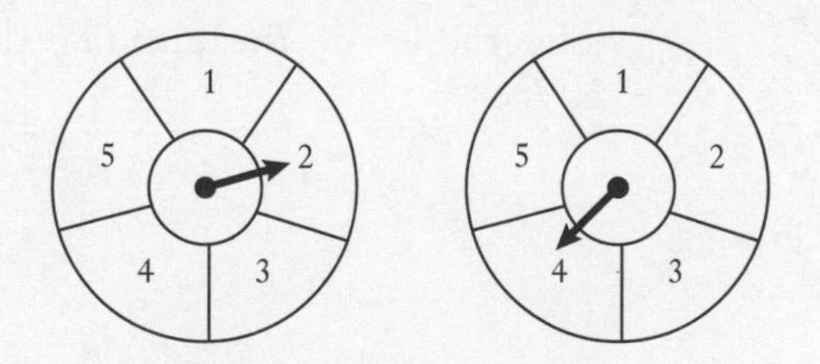

The two arrows are spun simultaneously.

Find the probability that the total of the resulting two numbers will be less than 6. — KU 2

DO NOT WRITE IN THIS MARGIN

KU	RE

7. Here are cards showing a number pattern:

Card 1: $\boxed{0 \times 1 - 1 \times (-3)}$

Card 2: $\boxed{1 \times 2 - 2 \times (-2)}$

Card 3: $\boxed{2 \times 3 - 3 \times (-1)}$

(*a*) Write down the number pattern that would appear on card 4. **1**

(*b*) Hence or otherwise find the expression that would appear on card *n* in its simplest form. **3**

8. The times (to the nearest 10 minutes) spent on homework by a group of pupils are shown below:

Time spent (min)	0	10	20	30	40	50	60
Frequency	1	7	11	22	19	8	2

Construct a cumulative frequency table and hence find the median time spent on homework for this group of pupils. **3**

9. The diagram shows the cross-section of a circular railway tunnel.

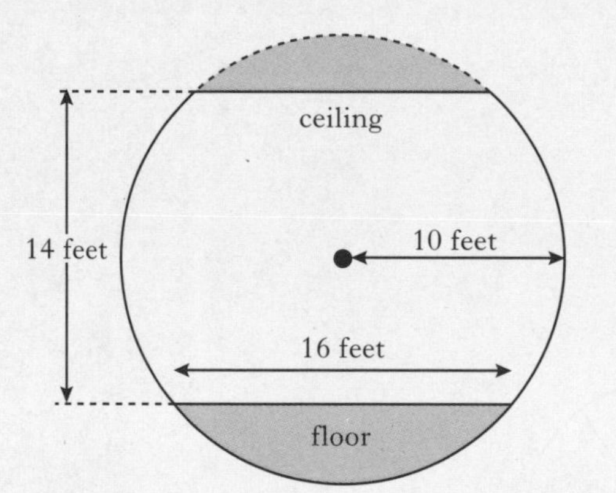

Horizontal ceiling and floor segments have been removed (shaded areas).

The height of the tunnel from floor to ceiling is 14 feet with the floor width being 16 feet.

The radius of the circular tunnel is 10 feet.

Calculate the width of the ceiling. **5**

10. $V = 4\sqrt{3} + 3\sqrt{A}$.

(*a*) Find the value of V when A = 27 expressing your answer as a surd in its simplest form. **3**

(*b*) Calculate the value of A when $V = 5\sqrt{3}$. **3**

DO NOT WRITE IN THIS MARGIN

KU	RE

11. (*a*) Mark plans to print leaflets advertising his gardening business. He has a budget of £100. Printing costs are 70p for 25 leaflets. In addition the printer charges a fixed administration fee of £51 on all orders.

How many leaflets can he afford to print? **2** (KU)

(*b*) A month later with a budget of £B he orders more leaflets. Printing costs are now c pence for 25 leaflets and the fixed administration fee is now £A.

Find a formula for T, the total number of leaflets he can afford to print. **3** (RE)

12. The diagram shows the rectangular end of an MP3 player. The area of the end is 1 cm^2.

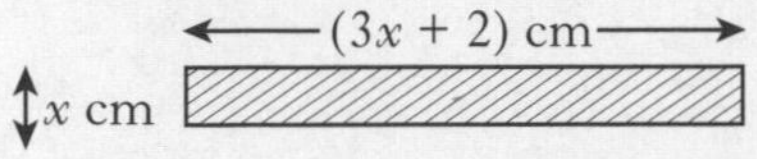

$(3x + 2)$ cm

x cm

Calculate the value of x, the thickness of the MP3 player. **5** (RE)

[End of question paper]

Mathematics | Standard Grade | Credit

Practice Papers
For SQA Exams

Exam D
Credit Level
Paper 2

You are allowed 1 hour, 20 minutes to complete this paper.

A calculator can be used.

Try to answer all of the questions in the time allowed, including all of your working.

Full marks will only be awarded where your answer includes any relevant working.

FORMULAE LIST

The roots of $ax^2 + bx + c = 0$ are $x = \dfrac{-b \pm \sqrt{(b^2 - 4ac)}}{2a}$

Sine Rule: $\dfrac{a}{\sin A} = \dfrac{b}{\sin B} = \dfrac{c}{\sin C}$

Cosine Rule: $a^2 = b^2 + c^2 - 2bc\cos A$ or $\cos A = \dfrac{b^2 + c^2 - a^2}{2bc}$

Area of a triangle: Area $= \dfrac{1}{2}ab\sin C$

Standard Deviation: $s = \sqrt{\dfrac{\Sigma(x - \overline{x})^2}{n-1}} = \sqrt{\dfrac{\Sigma x^2 - (\Sigma x)^2 / n}{n-1}}$, where n is the sample size.

DO NOT WRITE IN THIS MARGIN

KU | RE

1. The battery 'run time' of 7 laptops was measured. The times, measured to the nearest hour, were:

10, 11, 12, 12, 8, 10, 14.

Calculate the mean and standard deviation of these times. **4** (KU)

2. The diagram shows an example of a 'Golden Triangle'. Calculate the area of Golden Triangle ABC.

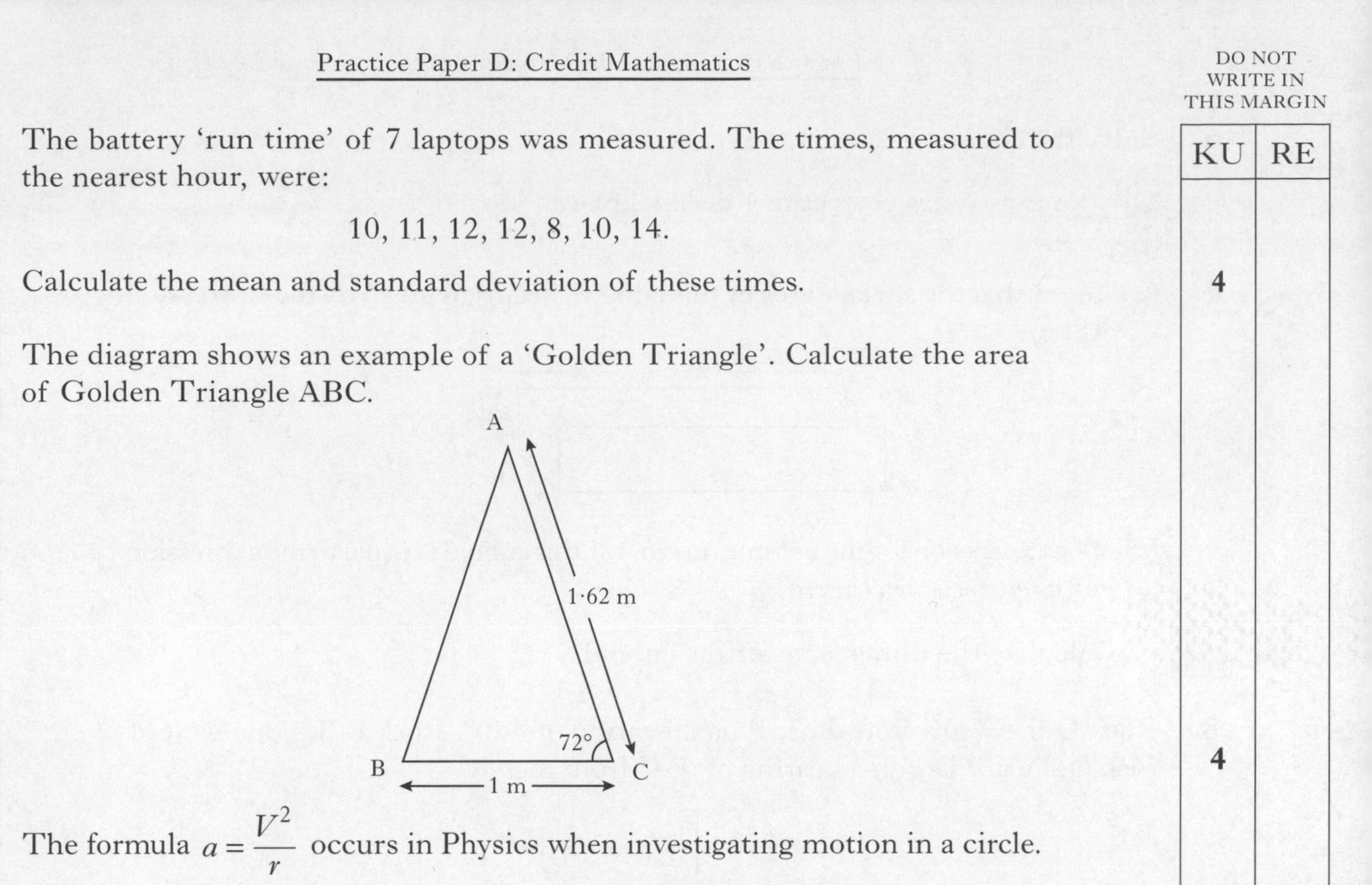

4 (KU)

3. The formula $a = \frac{V^2}{r}$ occurs in Physics when investigating motion in a circle.

Find the value of a when $V = 1{\cdot}2 \times 10^3$ and $r = 4 \times 10^8$. Give your answer in scientific notation. **3** (KU)

4. Pete arranged three sticks to form a triangle as shown.

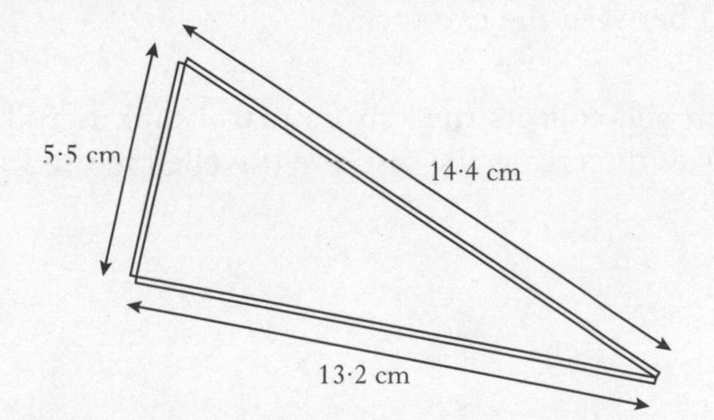

He thought the triangle was right-angled. Was he correct? Show your reasoning. **3** (RE)

5. The diagram shows the cross-section of the attic space of a house which is in the shape of an isosceles triangle.

There are three vertical roof supports with the longest central support 4 metres in height. The two smaller supports are each placed 2 metres from the edge of the floor. The whole floor measures 12 metres across.

Calculate the height of the two small supports. **4** (RE)

DO NOT WRITE IN THIS MARGIN

	KU	RE

6. Solve the equation $2x^2 + 3 = 6x$.

Give your answers correct to 1 decimal place. **3** (KU)

7. (*a*) Show that the surface area of this cuboid, in cm^2, is given by the expression $18x^2$. **1** (RE)

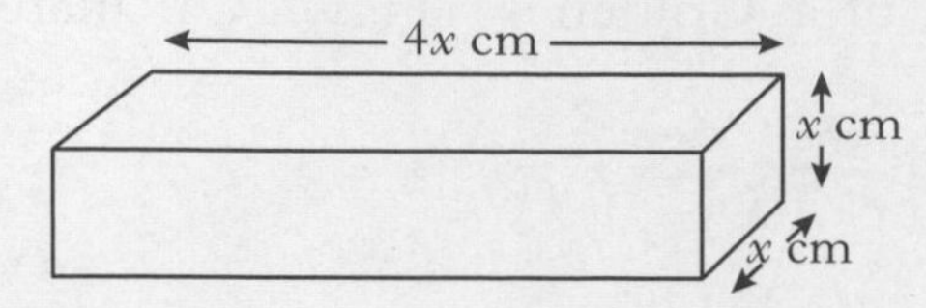

(*b*) The expression for the volume (in cm^3) of this cuboid is equal to the expression for its surface area (in cm^2).

Calculate the dimensions of the cuboid. **4** (RE)

8. Rock Q lies 7 km from Port P on a bearing of 120°. Rock R lies due west of rock Q. Port P lies on a bearing of 054° from rock R.

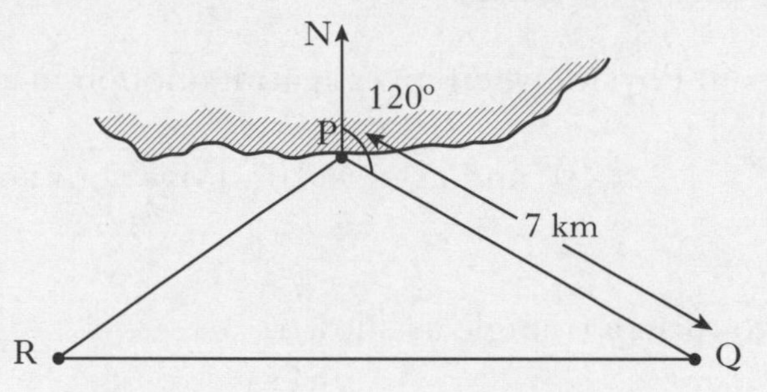

Calculate the distance RQ between the two rocks. **5** (KU)

9. Megan rents a van. When she collects the van its petrol tank is full. The amount of petrol used varies directly as the distance travelled in the van.

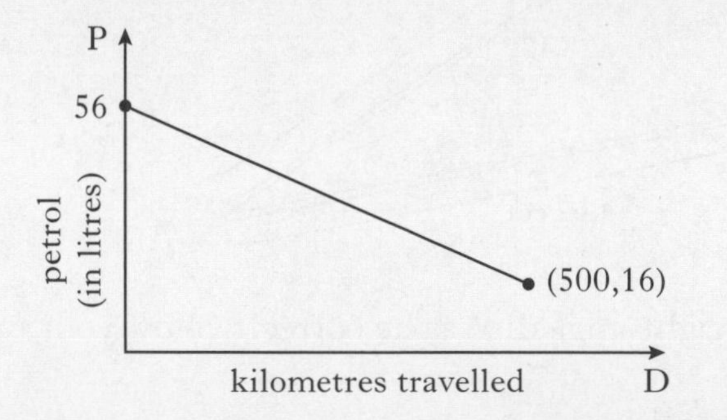

The graph shows the amount of petrol, P litres, left in the tank and the distance travelled, D km.

(*a*) How much petrol was in the tank initially? **1** (RE)

(*b*) Calculate the rate of petrol consumption in litres per 100 km. **3** (RE)

10. (*a*) Solve algebraically the equation $4\cos x° - 1 = 0 \quad 0 \leq x < 360$ **3** (KU)

(*b*) Hence write down the solution of the equation

$4\cos\left(\frac{x}{2}\right)^{\circ} - 1 = 0 \quad 0 \leq x < 360.$ **1** (RE)

DO NOT WRITE IN THIS MARGIN

KU	RE

11.

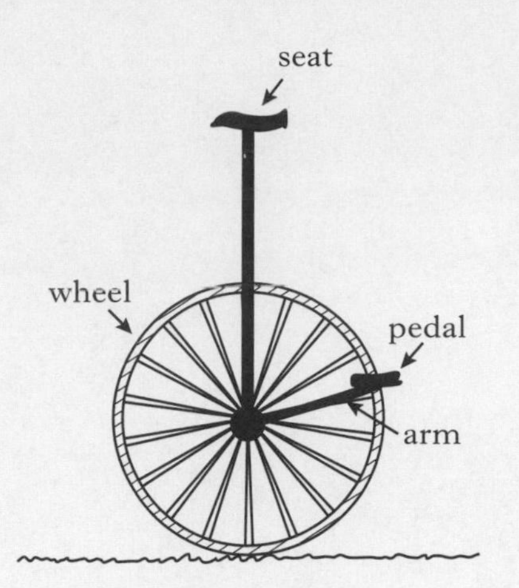

This diagram shows the various parts of a unicycle.

The diagram on the right represents this unicycle.

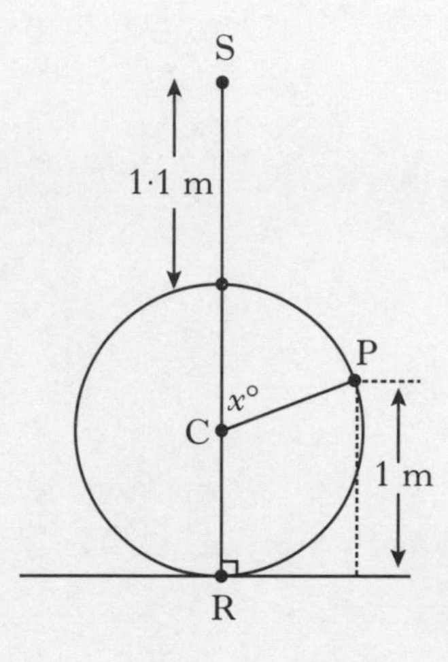

The seat, S, lies vertically above the centre, C, of the wheel.

The wheel has a radius of 0·75 m.

The seat, S, is 1·1 m above the top of the wheel.

The pedal, P, is 1 m above the horizontal road surface.

(*a*) Calculate the angle $x°$ which the arm of the pedal, CP, makes with the vertical. **4**

(*b*) Calculate how far, in metres, the pedal P must travel clockwise to reach position R vertically below the centre C. Give your answer correct to 3 significant figures. **4**

[End of question paper]

Practice Exam E

Mathematics | Standard Grade | Credit

Practice Papers
For SQA Exams

Exam E
Credit Level
Paper 1
Non-calculator

You are allowed 55 minutes to complete this paper.

Do **not** use a calculator.

Try to answer all of the questions in the time allowed, including all of your working.

Full marks will only be awarded where your answer includes any relevant working.

FORMULAE LIST

The roots of $ax^2 + bx + c = 0$ are $x = \dfrac{-b \pm \sqrt{\left(b^2 - 4ac\right)}}{2a}$

Sine Rule: $\dfrac{a}{\sin A} = \dfrac{b}{\sin B} = \dfrac{c}{\sin C}$

Cosine Rule: $a^2 = b^2 + c^2 - 2bc\cos A$ or $\cos A = \dfrac{b^2 + c^2 - a^2}{2bc}$

Area of a triangle: Area $= \dfrac{1}{2}ab\sin C$

Standard Deviation: $s = \sqrt{\dfrac{\Sigma(x - \overline{x})^2}{n-1}} = \dfrac{\sqrt{\Sigma x^2 - (\Sigma x)^2 / n}}{n-1}$, where n is the sample size.

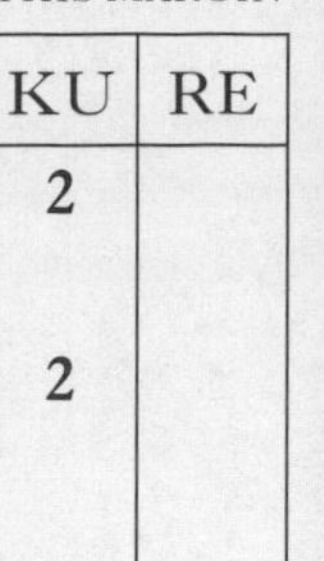

1. Evaluate $\frac{1}{9} \div 2\frac{1}{6}$. **2** (KU)

2. Evaluate $9{\cdot}9 - 7{\cdot}2 \div 30$. **2** (KU)

3. $a = \frac{1+3b}{bc}$.

Change the subject of this formula to b. **3** (KU)

4. Fergus rolls a pair of dice 90 times, hoping to get a total of either 4 or 6.

If the probability of this happening is $\frac{2}{9}$, how many times out of 90 rolls would you expect one of these totals to appear? **2** (RE)

5.

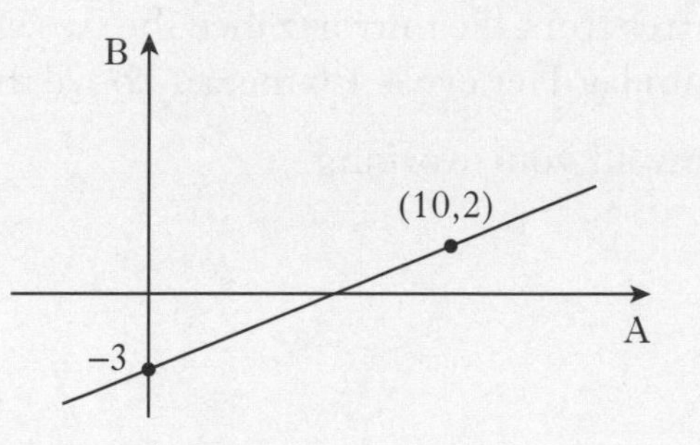

Find the equation of this straight line graph in terms of A and B. **4** (KU)

6. Remove brackets and simplify $(3 - 2a)(2 + a) - (3 - a^2)$. **3** (KU)

DO NOT WRITE IN THIS MARGIN

KU	RE

7. Chloe estimated the height of a lamp post to be more than 6 metres.

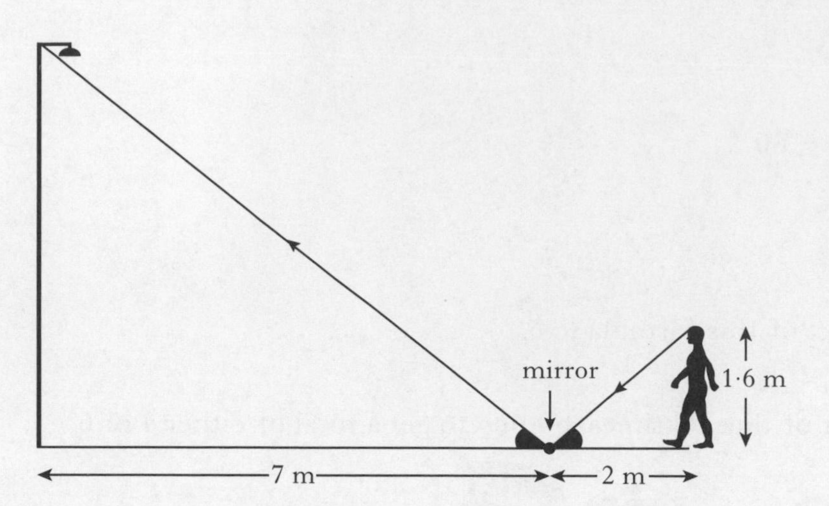

She now calculates the height using similar triangles. To do this she places a small mirror on the ground 7 metres from the base of the lamppost as shown.

She now walks back from the mirror until she can just see the top of the lamp post reflected in the mirror.

This happens when she is 2 metres from the mirror. Since the two shaded angles are equal the two triangles are similar. Her eye is 1·6 metres above the ground.

Was her estimate correct? Show all your working. — RE 3

8. Simplify $\sqrt{a} \times a^{3/2}$. — KU 2

9. (*a*) Keiran has saved up 30 coins. He has x twenty pences and y fifty pences. Write down an equation in x and y to illustrate this information. — KU 1

(*b*) A twenty pence coin weighs 5 g and a fifty pence coin weighs 8 g. When Keiran weighs his coins on the kitchen scales the total weight is 210 g. Write down another equation in x and y to illustrate this information. — KU 2

(*c*) How many of each coin does he have? Show your working. — RE 3

10. A rectangle has length $2x$ metres and breadth x metres.

It has a diagonal 100 metres in length.

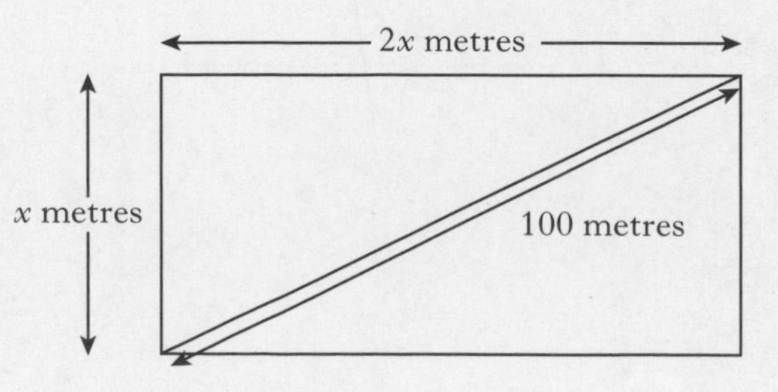

Calculate the value of x, giving your answer as a surd in simplest form. — RE 3

DO NOT WRITE IN THIS MARGIN

KU	RE

11. A relationship between A and B is given by the formula

$$A = \frac{1000}{\sqrt{B}}.$$

When B is multiplied by 4 what is the effect on A? **2**

12. Part of the graph $y = a\sin bx°$ is shown below.

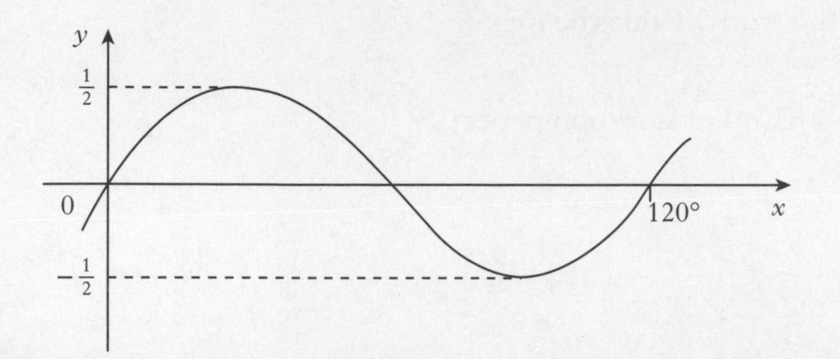

Write down the values of a and b. **2**

13. Archaeologists discovered a prehistoric circular enclosure with diameter 20 metres that was cut through by a roman road.

One edge of the road passes through the centre of the enclosure. The other edge forms a 16 metre long chord as shown in the diagram.

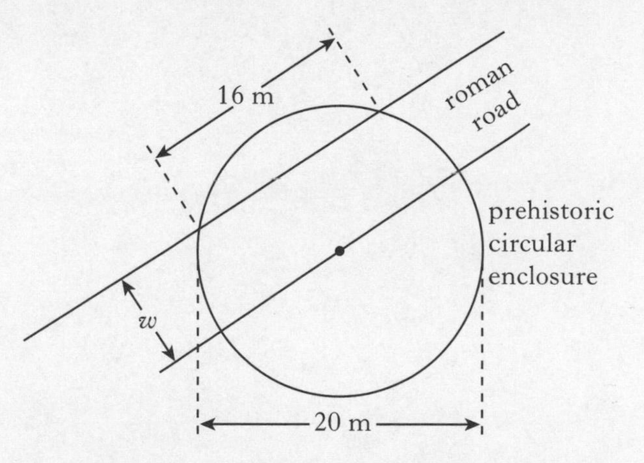

Calculate w, the width of the road, in metres. **4**

KU	RE

14. The nth term of a sequence is given by the formula

$$u_n = 2^n - 1.$$

(*a*) Find the 3rd term of this sequence. **1**

(*b*) There is a theorem that states:

If $2^n - 1$ is a prime number then n is prime also. It is known that 127 is a prime. By calculating a suitable value for n, show that this prime number is an example of the truth of this theorem. **2**

[End of question paper]

Mathematics | Standard Grade | Credit

Practice Papers
For SQA Exams

Exam E
Credit Level
Paper 2

You are allowed 1 hour, 20 minutes to complete this paper.

A calculator can be used.

Try to answer all of the questions in the time allowed, including all of your working.

Full marks will only be awarded where your answer includes any relevant working.

FORMULAE LIST

The roots of $ax^2 + bx + c = 0$ are $x = \dfrac{-b \pm \sqrt{(b^2 - 4ac)}}{2a}$

Sine Rule: $\dfrac{a}{\sin A} = \dfrac{b}{\sin B} = \dfrac{c}{\sin C}$

Cosine Rule: $a^2 = b^2 + c^2 - 2bc\cos A$ or $\cos A = \dfrac{b^2 + c^2 - a^2}{2bc}$

Area of a triangle: Area $= \dfrac{1}{2}ab\sin C$

Standard Deviation: $s = \sqrt{\dfrac{\Sigma(x - \bar{x})^2}{n-1}} = \sqrt{\dfrac{\Sigma x^2 - (\Sigma x)^2/n}{n-1}}$, where n is the sample size.

DO NOT WRITE IN THIS MARGIN

	KU	RE
1. Solve the equation $5x^2 + x - 1 = 0$. Give your answer correct to 2 significant figures.	4	
2. (*a*) Cameron measured the height of six Australian Mountain Ash trees which are reputedly the tallest of all flowering trees. The heights, in metres, were 92, 89, 88, 95, 96, 86. Find the mean and standard deviation for this data.	4	
(*b*) In a second location he again measured the heights of six Mountain Ash trees. At this second site the heights had a mean of 76 metres and a standard deviation of 15 metres. Make two valid comparisons between the heights of the trees at the first and second locations.		2
3. Kim purchased a plot of land at the start of 2011 for £15 500. She estimated that its value would appreciate by 6·5% each year. Assuming her estimate is correct what would the plot be worth at the start of 2014.	3	
4. Stephen has received a pay rise of 15% and now earns £15·64 per hour. What was his hourly rate of pay before the rise?	3	
5. Solve the inequality $\frac{x-3}{6} > \frac{2}{3}$	2	
6. A lawn sprinkler system waters an area in the shape of a sector of a circle with radius 20 metres. The angle at the sprinkler is 48°. Calculate the area of lawn covered by the sprinkler.	3	

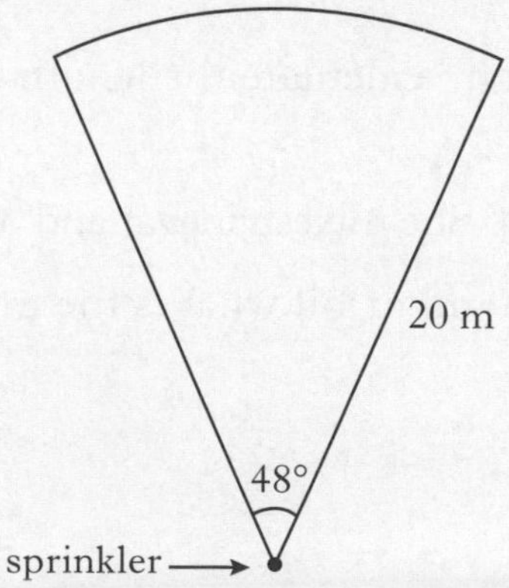

DO NOT WRITE IN THIS MARGIN

	KU	RE

7. Kirkcaldy is 11 km due East of Cowdenbeath. From Cowdenbeath the bearing of Leven is 066°. From Kirkcaldy the bearing of Leven is 048°.

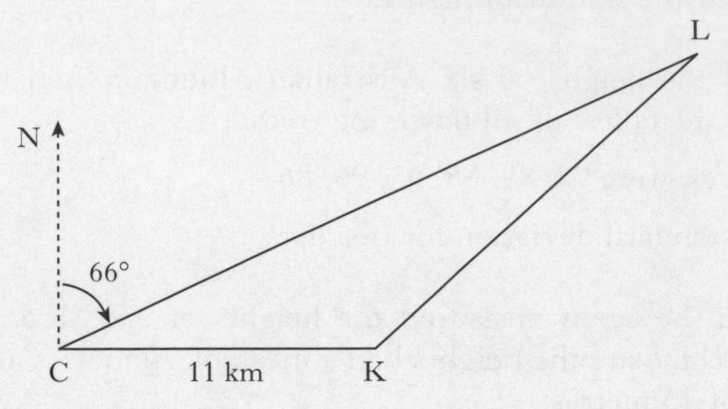

Calculate the distance between Cowdenbeath and Leven. **4**

8. Solve algebraically the equation

$3\tan x° + 7 = 0, \quad 0 \leq x < 360.$ **3**

9. The diagram shows a flat triangular metal machine component.

Side AB is 3 cm in length.

The angle between sides AB and AC is 110°.

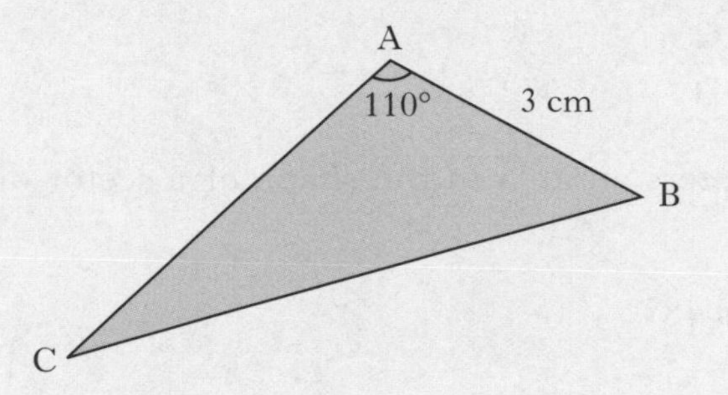

If the area of the component is $6{\cdot}3$ cm² calculate the length of side AC. **3**

10. Evie makes a vinaigrette salad dressing. She mixes vinegar and oil in the ratio 4:9.

If she has 248 ml of vinegar and 600 ml of oil what is the maximum volume of vinaigrette that she can make? **3**

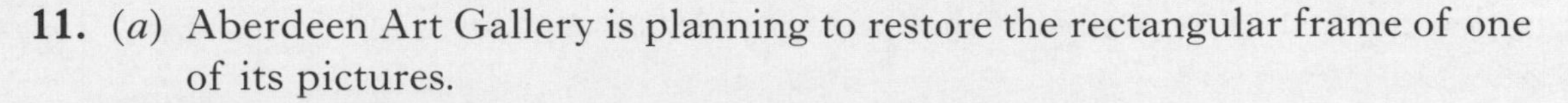

KU	RE

11. (*a*) Aberdeen Art Gallery is planning to restore the rectangular frame of one of its pictures.

Here is a diagram of the rectangular picture with its surrounding frame:

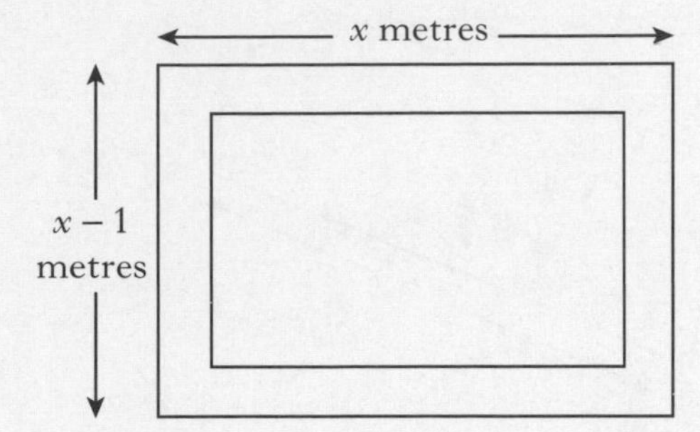

The outside dimensions of the frame are x metres by $(x - 1)$ metres.

The frame has a uniform width of 0·5 metres.

Show that the area, P m^2, of the picture is given by

$$P = x^2 - 3x + 2$$ **2**

(*b*) If the area of the picture is $\frac{3}{5}$ of the total area enclosed (picture plus frame), calculate the value of x. **4**

12.

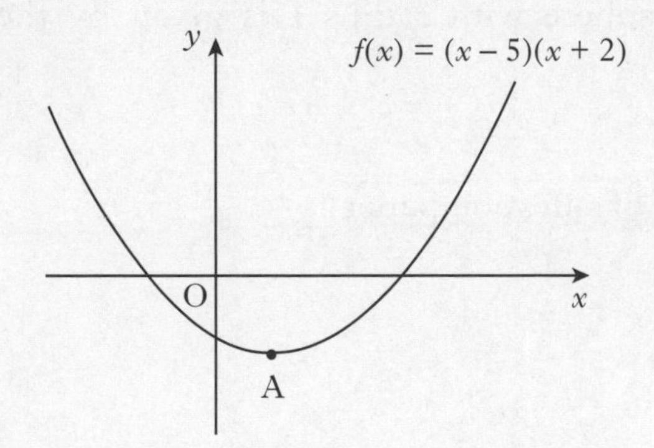

The diagram shows the graph of the function $f(x) = (x - 5)(x + 2)$.

Find the y-coordinate of the minimum turning point A on the graph. **4**

DO NOT WRITE IN THIS MARGIN

KU	RE

13. (*a*) A garden centre has 'row covers' that protect plants from frost.

They are shaped as half-cylinders 8 metres long and stand 2 metres tall at their highest point.

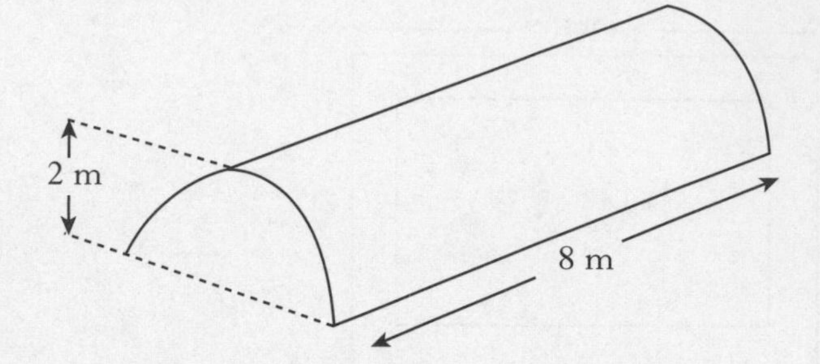

Calculate the volume of air enclosed by this 'row cover'. **2** (KU)

(*b*) The garden centre plans to replace each 'row cover' by a hemispherical cover that encloses the same volume of air.

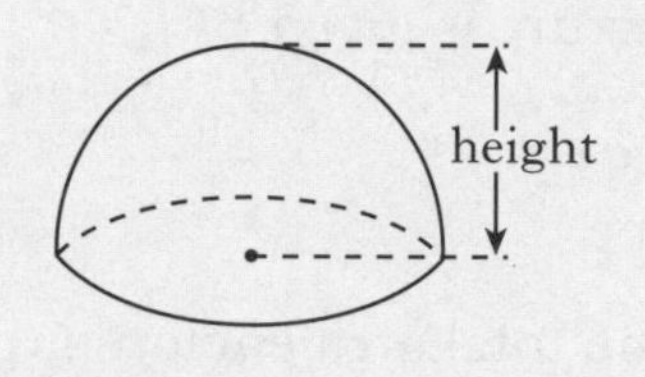

Calculate the height of this hemisphere.

[The volume of a hemisphere with radius r is given by the formula $V = \frac{2}{3}\pi r^3$.] **3** (RE)

[End of question paper]

Practice Exam F

Mathematics | Standard Grade | Credit

Practice Papers
For SQA Exams

Exam F
Credit Level
Paper 1
Non-calculator

You are allowed 55 minutes to complete this paper.

Do **not** use a calculator.

Try to answer all of the questions in the time allowed, including all of your working.

Full marks will only be awarded where your answer includes any relevant working.

FORMULAE LIST

The roots of $ax^2 + bx + c = 0$ are $x = \dfrac{-b \pm \sqrt{(b^2 - 4ac)}}{2a}$

Sine Rule: $\dfrac{a}{\sin A} = \dfrac{b}{\sin B} = \dfrac{c}{\sin C}$

Cosine Rule: $a^2 = b^2 + c^2 - 2bc \cos A$ or $\cos A = \dfrac{b^2 + c^2 - a^2}{2bc}$

Area of a triangle: $\text{Area} = \dfrac{1}{2} ab \sin C$

Standard Deviation: $s = \sqrt{\dfrac{\Sigma(x - \bar{x})^2}{n-1}} = \sqrt{\dfrac{\Sigma x^2 - (\Sigma x)^2 / n}{n-1}}$, where n is the sample size.

DO NOT WRITE IN THIS MARGIN

		KU	RE
1.	Evaluate $2\frac{2}{3} - 1\frac{4}{5}$.	2	
2.	Evaluate $82{\cdot}8 \div (1{\cdot}8 - 0{\cdot}9)$.	2	
3.	(*a*) Factorise $2x^2 + x - 6$.	1	
	(*b*) Hence simplify $\dfrac{2x+4}{2x^2+x-6}$.	2	
4.	Given that $f(x) = \dfrac{6}{3+x}$, evaluate $f(-5)$.	2	
5.	Find the equation of this straight line graph.	3	
6.	These two hexagonal nuts are mathematically similar in shape.		

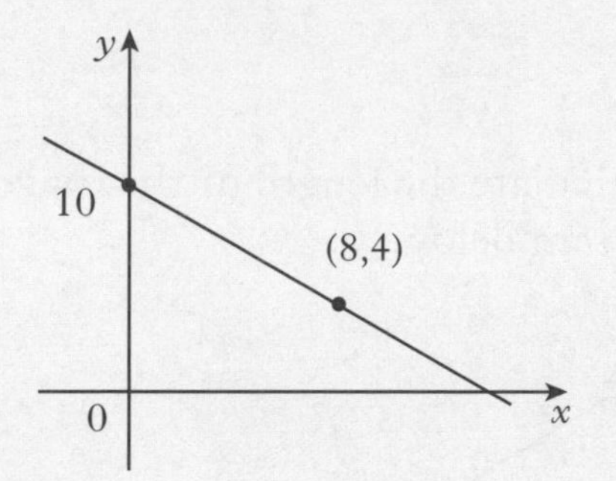

The thicknesses of the nuts are 10 mm for the larger one and 6 mm for the smaller one.

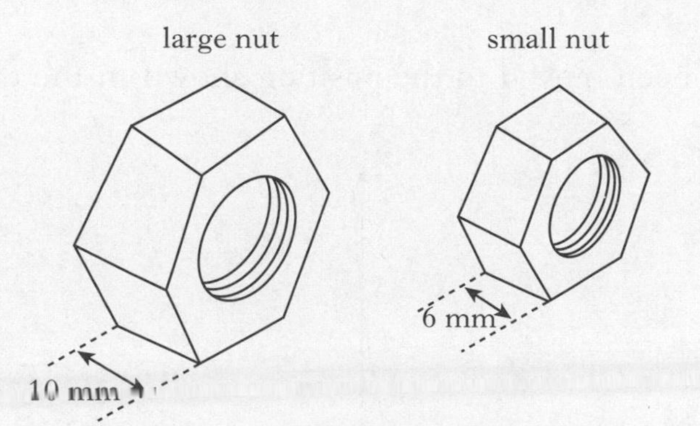

If the larger nut weights 15 g calculate the weight of the smaller nut. **4** (KU)

(You can assume that, for the two nuts, equal volumes have equal weights, i.e. the two nuts are made from the same metal).

	KU	RE

7. Solve the equation $3a - (2 - a) = 5a$. **3** (KU)

8.

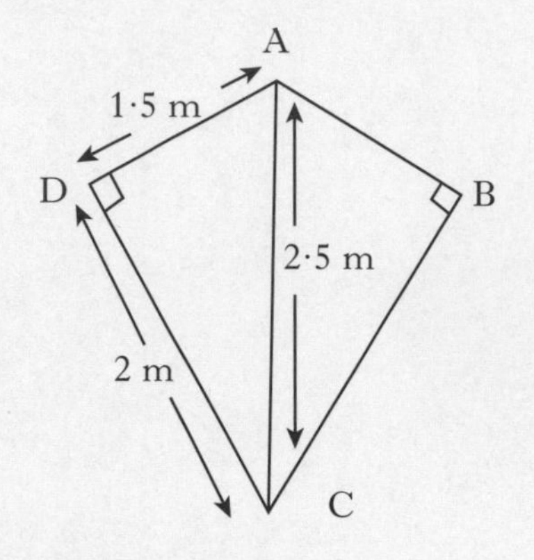

The diagram shows a kite ABCD with dimensions as shown.

The kite has right-angles at B and D.

(*a*) Calculate the area of the kite. **1** (KU)

(*b*) Use your answer to part (*a*) to calculate the length of the diagonal BD of kite ABCD as shown in the diagram below. **3** (RE)

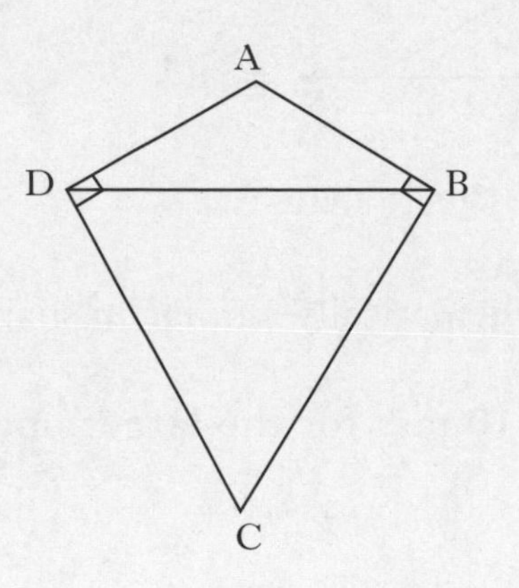

9. The graph of $y = x^2$ has been moved to the position shown in the diagram below.

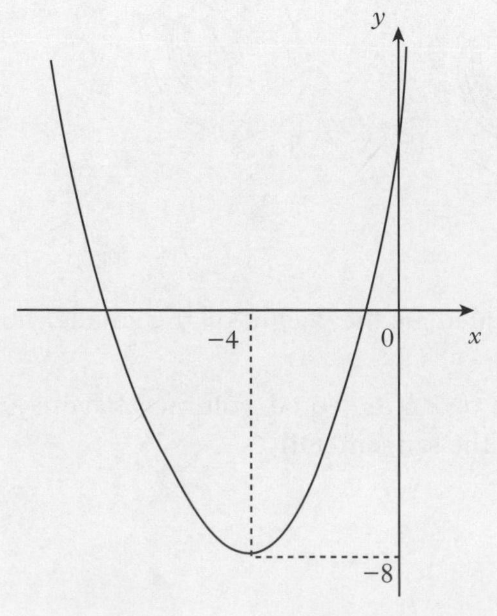

DO NOT WRITE IN THIS MARGIN

KU	RE

The equation of the graph in this new position is

$$y = (x + 4)^2 - 8.$$

Draw a sketch showing the graph with equation

$$y = (x - 3)^2 + 2.$$

RE: 2

10. Mia has 18 coins all of which are 10p-coins or 5p-coins. She has x 10 p-coins and y 5p-coins.

(*a*) Write down an equation in x and y to illustrate this information. KU: 1

(*b*) In total her coins are worth £1·25.

Write down another equation in x and y to illustrate this information. RE: 2

(*c*) How many 5p-coins does she have? RE: 3

11. (*a*) Mark downloads x tunes at a cost of £0·60 per download from Hometune's website. Write down an algebraic expression for the cost, in pence, of the downloads. RE: 1

(*b*) Hometune offers a package membership deal:

- Membership fee: only £8·00 per month.
- 10 free downloads to members per month.
- Special download rate for members: £0·45 per tune.

(i) Calculate the total cost to Mark of downloading 25 tunes during the course of a month if he were to become a member. KU: 1

(ii) Write down an algebraic expression for the total cost, in pence, for a member who downloads x tunes during the course of a month where x is greater than 10. RE: 2

(*c*) Find the minimum number of tunes Mark would have to download during the course of a month for membership to be a cheaper option than to download the tunes as a non-member.

Show all your working. RE: 3

[End of question paper]

Mathematics | Standard Grade | Credit

Practice Papers
For SQA Exams

Exam F
Credit Level
Paper 2

You are allowed 1 hour, 20 minutes to complete this paper.

A calculator can be used.

Try to answer all of the questions in the time allowed, including all of your working.

Full marks will only be awarded where your answer includes any relevant working.

FORMULAE LIST

The roots of $ax^2 + bx + c = 0$ are $x = \dfrac{-b \pm \sqrt{(b^2 - 4ac)}}{2a}$

Sine Rule: $\dfrac{a}{\sin A} = \dfrac{b}{\sin B} = \dfrac{c}{\sin C}$

Cosine Rule: $a^2 = b^2 + c^2 - 2bc\cos A$ or $\cos A = \dfrac{b^2 + c^2 - a^2}{2bc}$

Area of a triangle: Area $= \dfrac{1}{2}ab\sin C$

Standard Deviation: $s = \sqrt{\dfrac{\Sigma(x - \bar{x})^2}{n-1}} = \sqrt{\dfrac{\Sigma x^2 - (\Sigma x)^2/n}{n-1}}$, where n is the sample size.

DO NOT WRITE IN THIS MARGIN

KU	RE

1. A school purchased a computer for £273·70. The school later discovered that this price included V.A.T. at a rate of 15% for which they should not have been charged.

What should the school have been charged for the computer? **3**

2. Our sun, along with all the other stars in our galaxy, the Milky Way, orbits around a black hole that is at the centre of the galaxy.

The orbit is roughly circular with a radius of $2{\cdot}36 \times 10^{17}$ km. Calculate the circumference of the orbit.

Give your answer in scientific notation. **3**

3. (*a*) The average daily circulation figures, in thousands, over the course of 6 months for an Aberdeen local newspaper are:

27, 35, 24, 23, 29, 30.

Calculate the mean and standard deviation of this data. **3**

(*b*) The equivalent data over the same period for a local Dundee newspaper give a mean of 35 and a standard deviation of 1·2.

Make two valid comparsons between the circulation figures of the Aberdeen and the Dundee newspapers over the 6 months. **2**

4. A block of rubber in the shape of a cuboid with dimensions 8 cm × 3 cm × 4 cm is used to make a door wedge.

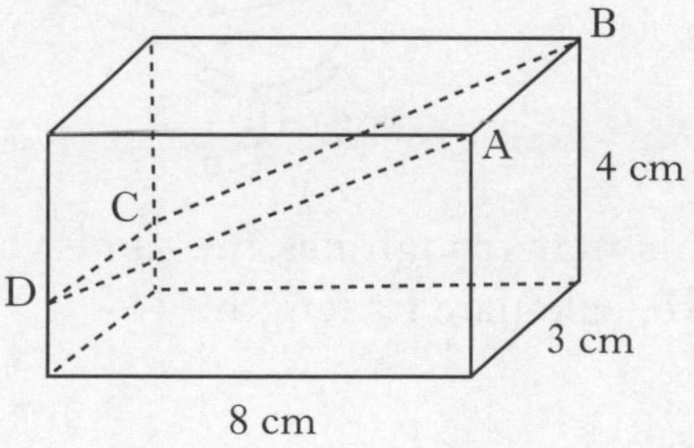

A slanting rectangular slice, ABCD, is made in the block through edge AB of the block.

This creates a wedge with height 4 cm at one end and 1 cm at the other end as shown in the diagram on the right.

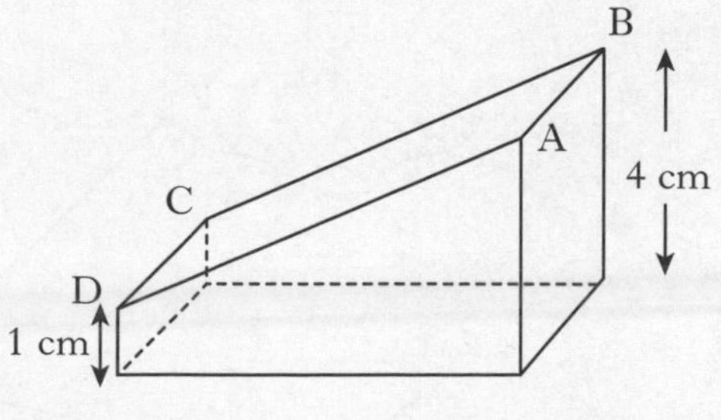

The wedge is sitting on a horizontal surface.

Calculate the angle the wedge makes with the horizontal.

Give your answer correct to 1 decimal place. **4**

DO NOT WRITE IN THIS MARGIN

	KU	RE
5. (*a*) Expand and simplify $(2y-1)(y-2)$.	1	
(*b*) Expand $x^{-\frac{1}{2}}(x-3)$.	2	
(*c*) Simplify, leaving your answer as a surd $\sqrt{108}-5\sqrt{3}$.	2	

6. (*a*) A cattle feeding trough is in the shape of a prism as shown in the diagram. It has length 4·2 m with the area of each end 1·6 m^2.

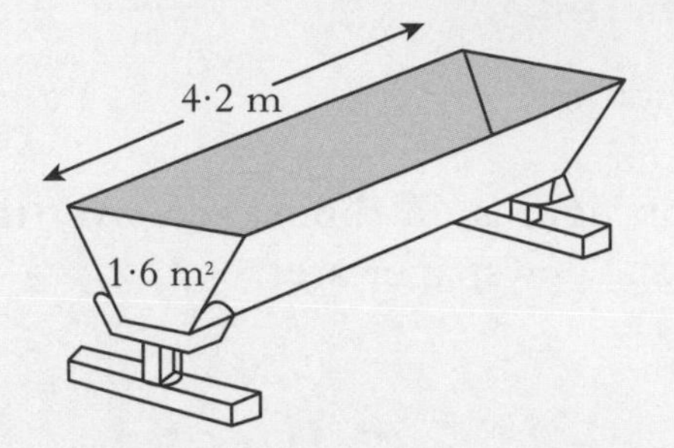

Calculate the volume of the trough. **1** (KU)

(*b*) An alternative design for the trough is in the shape of half a cylinder. For this design each end is a semi circle with diameter 2 metres.

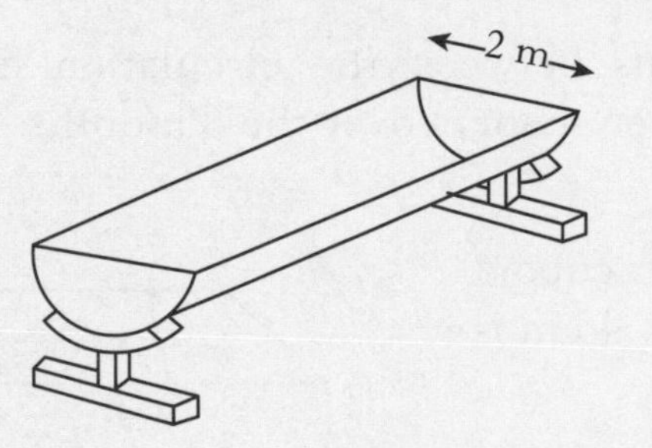

If this new trough has the same volume as the trough design in part (*a*) above, calculate its length. **4** (RE)

7.

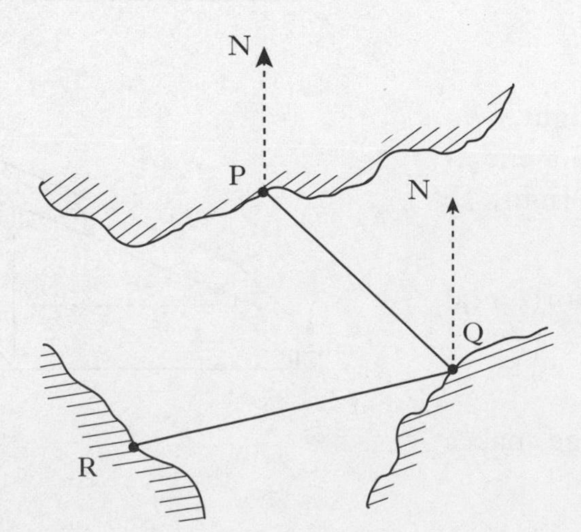

(*a*) Port Q lies on a bearing of 136° from Port P. From Port Q, the bearing of Port R is 258°.

Calculate the size of angle PQR. **2** (RE)

DO NOT WRITE IN THIS MARGIN

KU	RE

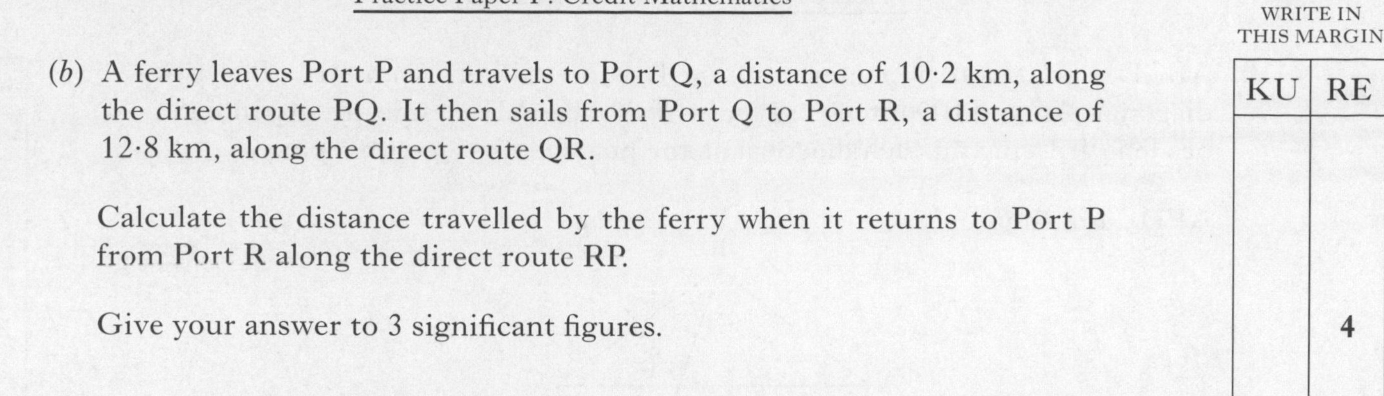

(*b*) A ferry leaves Port P and travels to Port Q, a distance of 10·2 km, along the direct route PQ. It then sails from Port Q to Port R, a distance of 12·8 km, along the direct route QR.

Calculate the distance travelled by the ferry when it returns to Port P from Port R along the direct route RP.

Give your answer to 3 significant figures. **4** (RE)

8. On this 1 square × 2 squares grid a total of 3 different rectangles can be drawn

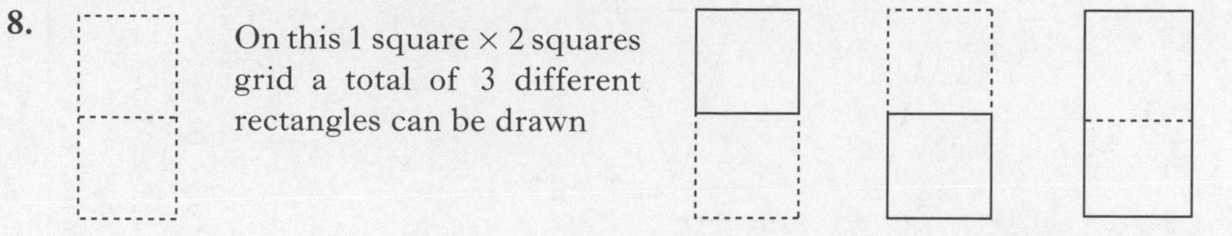

3 different rectangles

The number of different rectangles, r, that can be drawn on a grid of squares of size n squares × 2 squares is given by the formula

$$r = \frac{3}{2}n(n+1)$$

(*a*) How many different rectangles can be drawn on a grid of squares of size 3 squares × 2 squares? **2** (KU)

(*b*) On a grid of squares of size n squares × 2 squares it is possible to draw 198 different rectangles. Show that for this grid $n^2 + n - 132 = 0$. **2** (RE)

(*c*) Hence find the size of the grid of squares. **3** (RE)

9. The Large Hadron Collider has a circular tunnel with a radius of 4·3 km. In the tunnel protons are accelerated to nearly the speed of light. At this speed they travel 15 km in 51 microseconds (a microsecond is one millionth of a second). In the tunnel there is a detector, Alice, at point A and a detector, LHC 'B', at point B and they form an angle of 96° at the centre of the circle as shown.

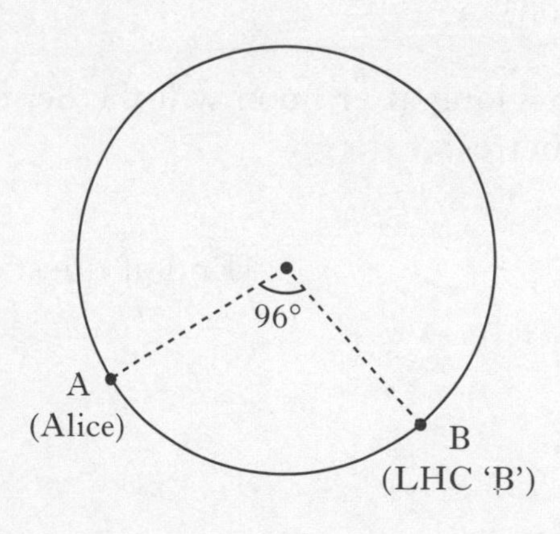

Calculate the time, in microseconds, a proton takes at this speed to travel anti-clockwise from Alice to LHC 'B'. **4** (RE)

DO NOT WRITE IN THIS MARGIN

KU	RE

10. ABCDE is a regular pentagon. Line L is an axis of symmetry. The two diagonals AC and BD intersect at point F as shown. Each side of the pentagon has length 1 cm and each diagonal of the pentagon has length x cm.

AFDE is a rhombus.

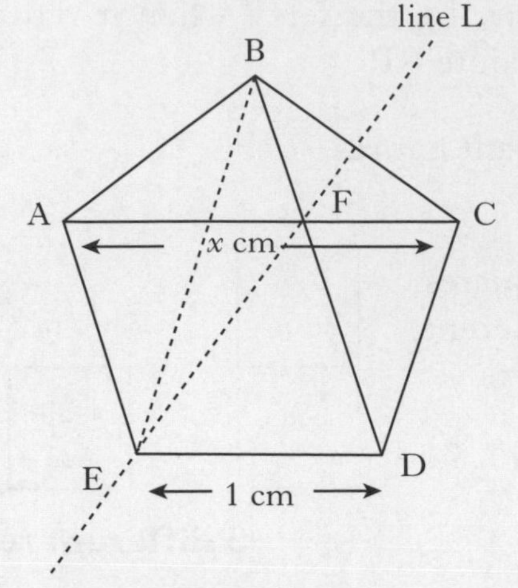

(*a*) Write down an expression for the length FC, in cm. **1**

(*b*) Triangle DFC is similar to triangle BDE. Use this information to show that $x^2 - x - 1 = 0$. **3**

11. The depth, d metres, of water in a harbour starting at noon is given by the formula

$d = 7–3 \sin (30h)°$, where h is the number of hours after noon.

(*a*) Calculate the depth of the water in the harbour at 3 pm (3 hours after noon). **2**

(*b*) How long after noon will the depth of the water in the harbour first be 5 metres? **4**

[End of question paper]

Worked Answers

Practice Exam D: Paper 1 Worked Answers

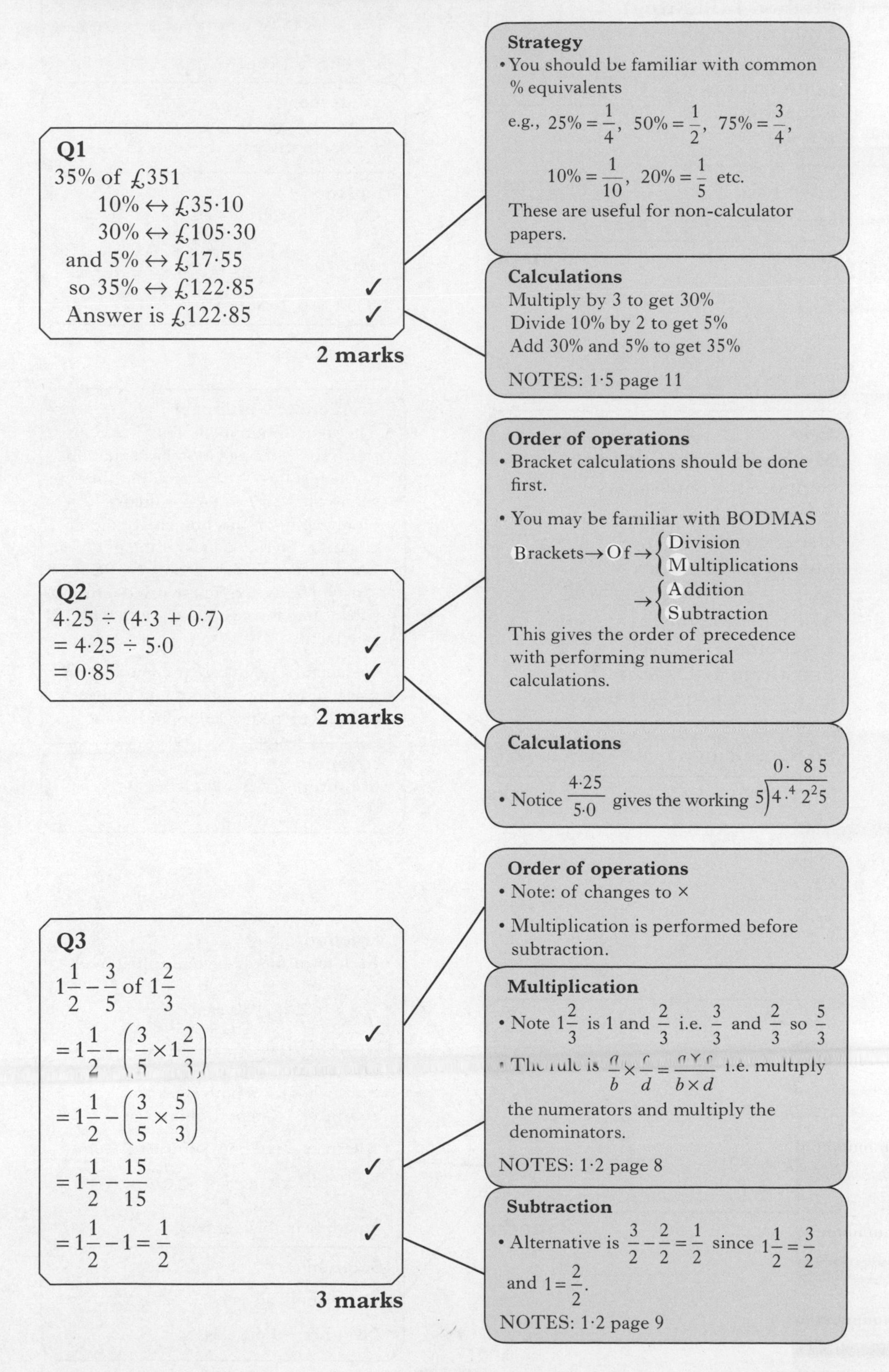

Q1

35% of £351

$10\% \leftrightarrow £35{\cdot}10$

$30\% \leftrightarrow £105{\cdot}30$

and $5\% \leftrightarrow £17{\cdot}55$

so $35\% \leftrightarrow £122{\cdot}85$ ✓

Answer is £122·85 ✓

2 marks

Strategy
- You should be familiar with common % equivalents

e.g., $25\% = \frac{1}{4}$, $50\% = \frac{1}{2}$, $75\% = \frac{3}{4}$,

$10\% = \frac{1}{10}$, $20\% = \frac{1}{5}$ etc.

These are useful for non-calculator papers.

Calculations

Multiply by 3 to get 30%

Divide 10% by 2 to get 5%

Add 30% and 5% to get 35%

NOTES: 1·5 page 11

Q2

$4{\cdot}25 \div (4{\cdot}3 + 0{\cdot}7)$

$= 4{\cdot}25 \div 5{\cdot}0$ ✓

$= 0{\cdot}85$ ✓

2 marks

Order of operations
- Bracket calculations should be done first.
- You may be familiar with BODMAS

Brackets → Of → {Division, Multiplications} → {Addition, Subtraction}

This gives the order of precedence with performing numerical calculations.

Calculations
- Notice $\frac{4{\cdot}25}{5{\cdot}0}$ gives the working $\begin{array}{r} 0{\cdot}\ 8\ 5 \\ 5\overline{)4{\cdot}^4 2^2 5} \end{array}$

Q3

$1\frac{1}{2} - \frac{3}{5}$ of $1\frac{2}{3}$

$= 1\frac{1}{2} - \left(\frac{3}{5} \times 1\frac{2}{3}\right)$ ✓

$= 1\frac{1}{2} - \left(\frac{3}{5} \times \frac{5}{3}\right)$

$= 1\frac{1}{2} - \frac{15}{15}$ ✓

$= 1\frac{1}{2} - 1 = \frac{1}{2}$ ✓

3 marks

Order of operations
- Note: of changes to ×
- Multiplication is performed before subtraction.

Multiplication
- Note $1\frac{2}{3}$ is 1 and $\frac{2}{3}$ i.e. $\frac{3}{3}$ and $\frac{2}{3}$ so $\frac{5}{3}$
- The rule is $\frac{a}{b} \times \frac{c}{d} = \frac{a \times c}{b \times d}$ i.e. multiply the numerators and multiply the denominators.

NOTES: 1·2 page 8

Subtraction
- Alternative is $\frac{3}{2} - \frac{2}{2} = \frac{1}{2}$ since $1\frac{1}{2} = \frac{3}{2}$ and $1 = \frac{2}{2}$.

NOTES: 1·2 page 9

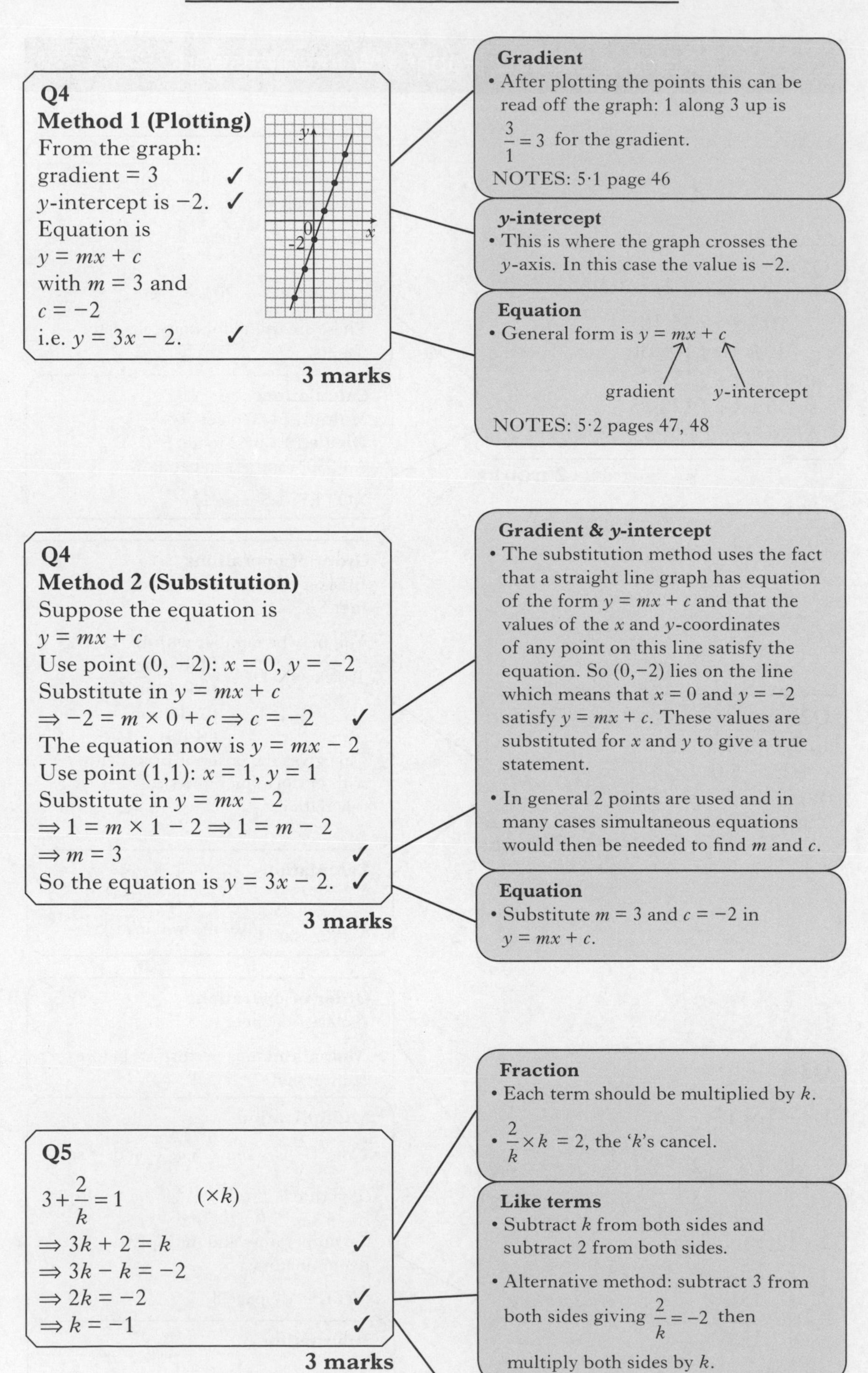

Q4
Method 1 (Plotting)
From the graph:
gradient = 3 ✓
y-intercept is -2. ✓
Equation is
$y = mx + c$
with $m = 3$ and
$c = -2$
i.e. $y = 3x - 2$. ✓

3 marks

Gradient
- After plotting the points this can be read off the graph: 1 along 3 up is $\frac{3}{1} = 3$ for the gradient.

NOTES: 5·1 page 46

***y*-intercept**
- This is where the graph crosses the y-axis. In this case the value is -2.

Equation
- General form is $y = mx + c$

gradient y-intercept

NOTES: 5·2 pages 47, 48

Q4
Method 2 (Substitution)
Suppose the equation is
$y = mx + c$
Use point $(0, -2)$: $x = 0$, $y = -2$
Substitute in $y = mx + c$
$\Rightarrow -2 = m \times 0 + c \Rightarrow c = -2$ ✓
The equation now is $y = mx - 2$
Use point $(1,1)$: $x = 1$, $y = 1$
Substitute in $y = mx - 2$
$\Rightarrow 1 = m \times 1 - 2 \Rightarrow 1 = m - 2$
$\Rightarrow m = 3$ ✓
So the equation is $y = 3x - 2$. ✓

3 marks

Gradient & *y*-intercept
- The substitution method uses the fact that a straight line graph has equation of the form $y = mx + c$ and that the values of the x and y-coordinates of any point on this line satisfy the equation. So $(0,-2)$ lies on the line which means that $x = 0$ and $y = -2$ satisfy $y = mx + c$. These values are substituted for x and y to give a true statement.
- In general 2 points are used and in many cases simultaneous equations would then be needed to find m and c.

Equation
- Substitute $m = 3$ and $c = -2$ in $y = mx + c$.

Q5
$3 + \frac{2}{k} = 1$ $(\times k)$
$\Rightarrow 3k + 2 = k$ ✓
$\Rightarrow 3k - k = -2$
$\Rightarrow 2k = -2$ ✓
$\Rightarrow k = -1$ ✓

3 marks

Fraction
- Each term should be multiplied by k.
- $\frac{2}{k} \times k = 2$, the 'k's cancel.

Like terms
- Subtract k from both sides and subtract 2 from both sides.
- Alternative method: subtract 3 from both sides giving $\frac{2}{k} = -2$ then multiply both sides by k.

Solution
- Divide both sides by 2: $\frac{2k}{2} = \frac{-2}{2}$

NOTES: 4·4 page 34

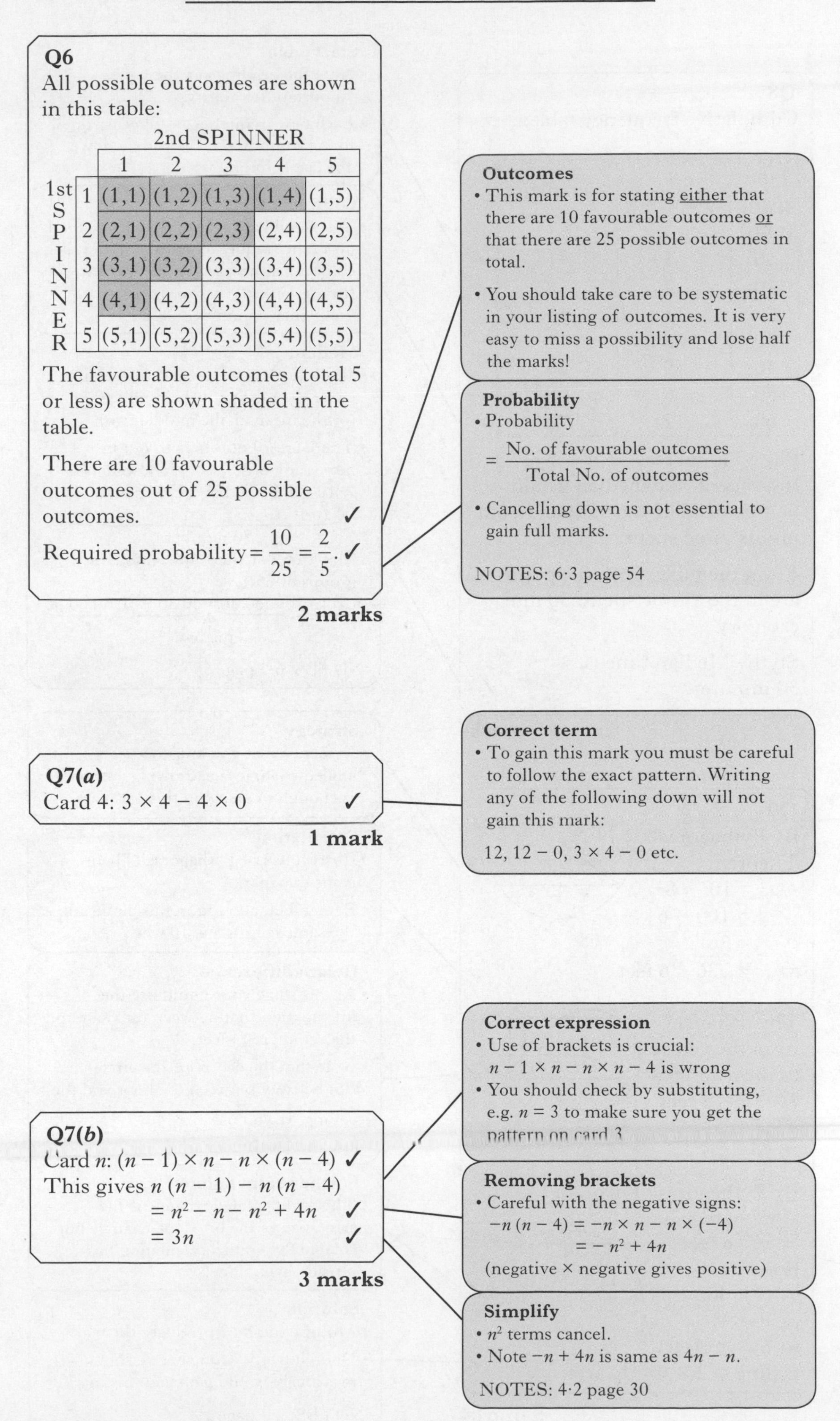

Q6

All possible outcomes are shown in this table:

1st SPINNER \ 2nd SPINNER	1	2	3	4	5
1	(1,1)	(1,2)	(1,3)	(1,4)	(1,5)
2	(2,1)	(2,2)	(2,3)	(2,4)	(2,5)
3	(3,1)	(3,2)	(3,3)	(3,4)	(3,5)
4	(4,1)	(4,2)	(4,3)	(4,4)	(4,5)
5	(5,1)	(5,2)	(5,3)	(5,4)	(5,5)

The favourable outcomes (total 5 or less) are shown shaded in the table.

There are 10 favourable outcomes out of 25 possible outcomes. ✓

Required probability $= \frac{10}{25} = \frac{2}{5}$. ✓

2 marks

Outcomes
- This mark is for stating <u>either</u> that there are 10 favourable outcomes <u>or</u> that there are 25 possible outcomes in total.
- You should take care to be systematic in your listing of outcomes. It is very easy to miss a possibility and lose half the marks!

Probability
- Probability

$$= \frac{\text{No. of favourable outcomes}}{\text{Total No. of outcomes}}$$

- Cancelling down is not essential to gain full marks.

NOTES: 6·3 page 54

Q7(*a*)

Card 4: $3 \times 4 - 4 \times 0$ ✓

1 mark

Correct term
- To gain this mark you must be careful to follow the exact pattern. Writing any of the following down will not gain this mark:

12, $12 - 0$, $3 \times 4 - 0$ etc.

Q7(*b*)

Card n: $(n-1) \times n - n \times (n-4)$ ✓

This gives $n(n-1) - n(n-4)$

$= n^2 - n - n^2 + 4n$ ✓

$= 3n$ ✓

3 marks

Correct expression
- Use of brackets is crucial: $n - 1 \times n - n \times n - 4$ is wrong
- You should check by substituting, e.g. $n = 3$ to make sure you get the pattern on card 3

Removing brackets
- Careful with the negative signs:

$-n(n-4) = -n \times n - n \times (-4)$

$= -n^2 + 4n$

(negative × negative gives positive)

Simplify
- n^2 terms cancel.
- Note $-n + 4n$ is same as $4n - n$.

NOTES: 4·2 page 30

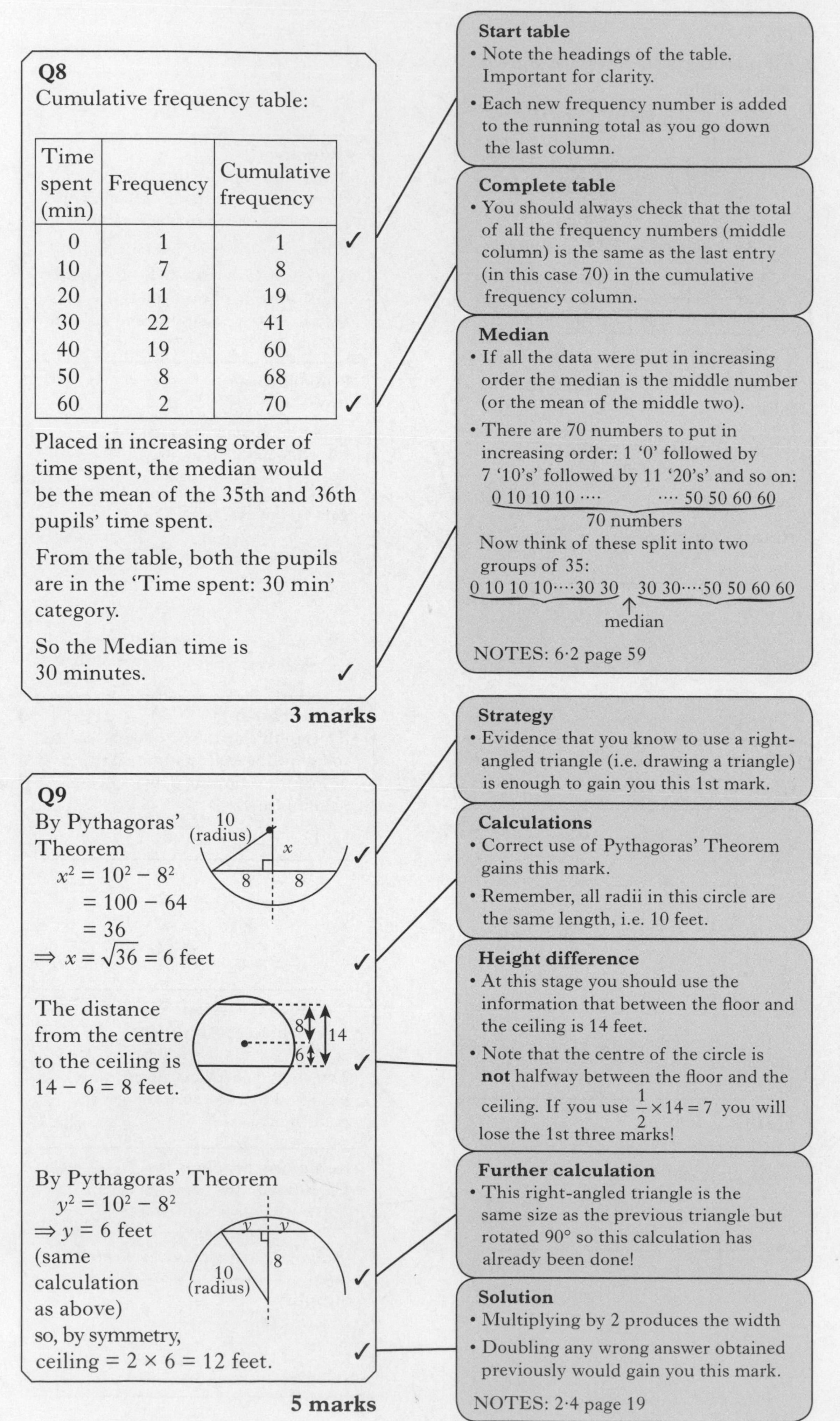

Q8

Cumulative frequency table:

Time spent (min)	Frequency	Cumulative frequency
0	1	1
10	7	8
20	11	19
30	22	41
40	19	60
50	8	68
60	2	70

✓ ✓

Placed in increasing order of time spent, the median would be the mean of the 35th and 36th pupils' time spent.

From the table, both the pupils are in the 'Time spent: 30 min' category.

So the Median time is 30 minutes. ✓

3 marks

Start table
- Note the headings of the table. Important for clarity.
- Each new frequency number is added to the running total as you go down the last column.

Complete table
- You should always check that the total of all the frequency numbers (middle column) is the same as the last entry (in this case 70) in the cumulative frequency column.

Median
- If all the data were put in increasing order the median is the middle number (or the mean of the middle two).
- There are 70 numbers to put in increasing order: 1 '0' followed by 7 '10's' followed by 11 '20's' and so on:

0 10 10 10 ···· ···· 50 50 60 60 (70 numbers)

Now think of these split into two groups of 35:

0 10 10 10····30 30 | 30 30····50 50 60 60 (↑ median)

NOTES: 6·2 page 59

Q9

By Pythagoras' Theorem

$x^2 = 10^2 - 8^2$
$= 100 - 64$
$= 36$
$\Rightarrow x = \sqrt{36} = 6$ feet ✓ ✓

The distance from the centre to the ceiling is $14 - 6 = 8$ feet. ✓

By Pythagoras' Theorem

$y^2 = 10^2 - 8^2$
$\Rightarrow y = 6$ feet
(same calculation as above)

✓

so, by symmetry, ceiling $= 2 \times 6 = 12$ feet. ✓

5 marks

Strategy
- Evidence that you know to use a right-angled triangle (i.e. drawing a triangle) is enough to gain you this 1st mark.

Calculations
- Correct use of Pythagoras' Theorem gains this mark.
- Remember, all radii in this circle are the same length, i.e. 10 feet.

Height difference
- At this stage you should use the information that between the floor and the ceiling is 14 feet.
- Note that the centre of the circle is **not** halfway between the floor and the ceiling. If you use $\frac{1}{2} \times 14 = 7$ you will lose the 1st three marks!

Further calculation
- This right-angled triangle is the same size as the previous triangle but rotated 90° so this calculation has already been done!

Solution
- Multiplying by 2 produces the width
- Doubling any wrong answer obtained previously would gain you this mark.

NOTES: 2·4 page 19

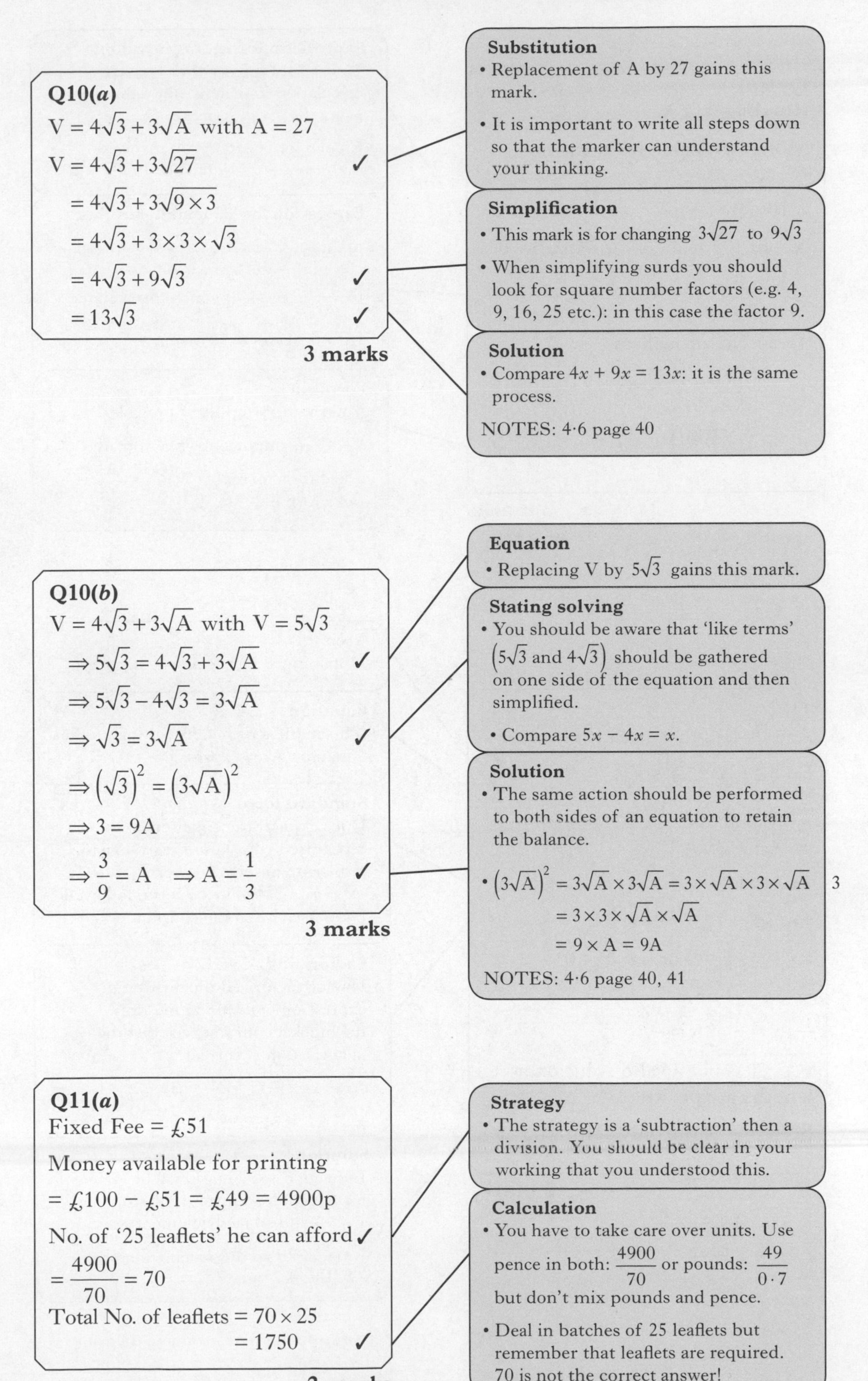

Q10(*a*)

$V = 4\sqrt{3} + 3\sqrt{A}$ with $A = 27$

$V = 4\sqrt{3} + 3\sqrt{27}$ ✓

$= 4\sqrt{3} + 3\sqrt{9 \times 3}$

$= 4\sqrt{3} + 3 \times 3 \times \sqrt{3}$

$= 4\sqrt{3} + 9\sqrt{3}$ ✓

$= 13\sqrt{3}$ ✓

3 marks

Substitution
- Replacement of A by 27 gains this mark.
- It is important to write all steps down so that the marker can understand your thinking.

Simplification
- This mark is for changing $3\sqrt{27}$ to $9\sqrt{3}$
- When simplifying surds you should look for square number factors (e.g. 4, 9, 16, 25 etc.): in this case the factor 9.

Solution
- Compare $4x + 9x = 13x$: it is the same process.

NOTES: 4·6 page 40

Q10(*b*)

$V = 4\sqrt{3} + 3\sqrt{A}$ with $V = 5\sqrt{3}$

$\Rightarrow 5\sqrt{3} = 4\sqrt{3} + 3\sqrt{A}$ ✓

$\Rightarrow 5\sqrt{3} - 4\sqrt{3} = 3\sqrt{A}$

$\Rightarrow \sqrt{3} = 3\sqrt{A}$ ✓

$\Rightarrow \left(\sqrt{3}\right)^2 = \left(3\sqrt{A}\right)^2$

$\Rightarrow 3 = 9A$

$\Rightarrow \frac{3}{9} = A \quad \Rightarrow A = \frac{1}{3}$ ✓

3 marks

Equation
- Replacing V by $5\sqrt{3}$ gains this mark.

Stating solving
- You should be aware that 'like terms' $\left(5\sqrt{3} \text{ and } 4\sqrt{3}\right)$ should be gathered on one side of the equation and then simplified.
- Compare $5x - 4x = x$.

Solution
- The same action should be performed to both sides of an equation to retain the balance.
- $\left(3\sqrt{A}\right)^2 = 3\sqrt{A} \times 3\sqrt{A} = 3 \times \sqrt{A} \times 3 \times \sqrt{A}$
  $= 3 \times 3 \times \sqrt{A} \times \sqrt{A}$
  $= 9 \times A = 9A$

NOTES: 4·6 page 40, 41

Q11(*a*)

Fixed Fee = £51

Money available for printing

= £100 − £51 = £49 = 4900p

No. of '25 leaflets' he can afford ✓

$= \frac{4900}{70} = 70$

Total No. of leaflets $= 70 \times 25$

$= 1750$ ✓

2 marks

Strategy
- The strategy is a 'subtraction' then a division. You should be clear in your working that you understood this.

Calculation
- You have to take care over units. Use pence in both: $\frac{4900}{70}$ or pounds: $\frac{49}{0 \cdot 7}$ but don't mix pounds and pence.
- Deal in batches of 25 leaflets but remember that leaflets are required. 70 is not the correct answer!

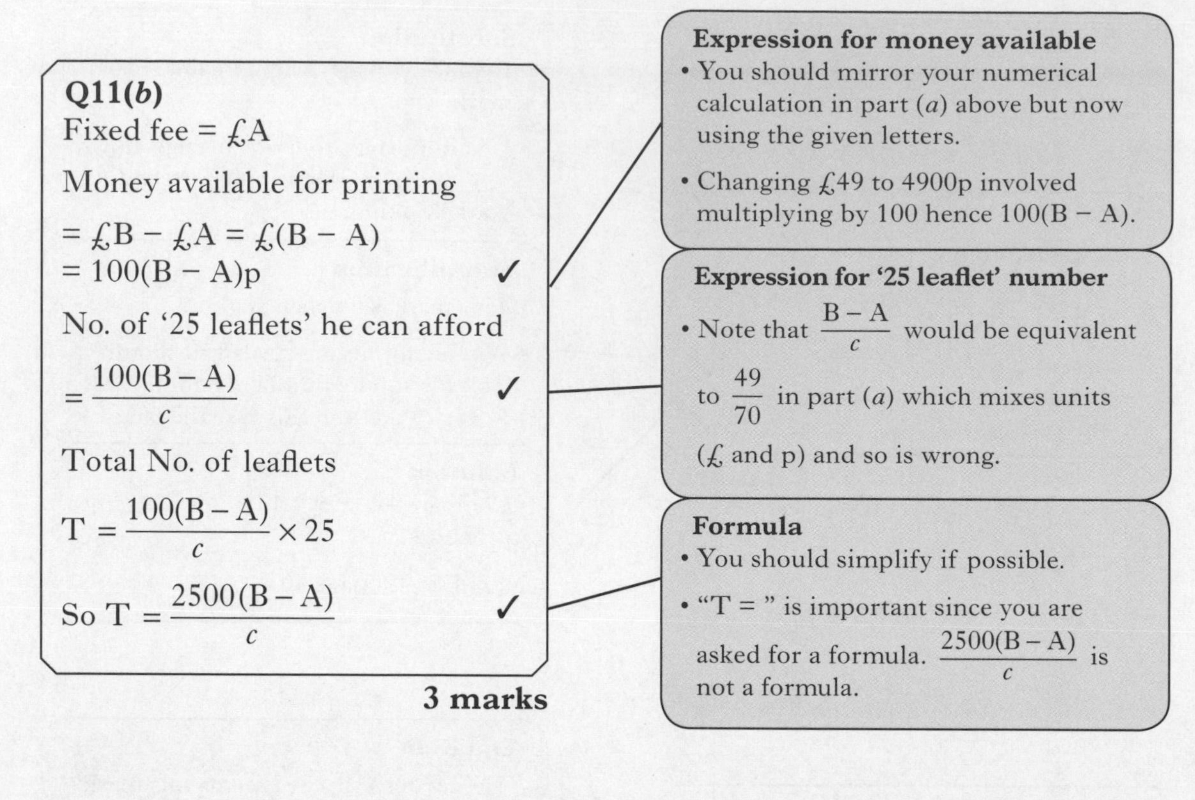

Q11(*b*)

Fixed fee = £A

Money available for printing

= £B − £A = £(B − A)

= 100(B − A)p ✓

No. of '25 leaflets' he can afford

$= \frac{100(B-A)}{c}$ ✓

Total No. of leaflets

$T = \frac{100(B-A)}{c} \times 25$

So $T = \frac{2500(B-A)}{c}$ ✓

3 marks

Expression for money available
- You should mirror your numerical calculation in part (*a*) above but now using the given letters.
- Changing £49 to 4900p involved multiplying by 100 hence 100(B − A).

Expression for '25 leaflet' number
- Note that $\frac{B-A}{c}$ would be equivalent to $\frac{49}{70}$ in part (*a*) which mixes units (£ and p) and so is wrong.

Formula
- You should simplify if possible.
- "T = " is important since you are asked for a formula. $\frac{2500(B-A)}{c}$ is not a formula.

Q12

Area = length × breadth ✓

So $1 = (3x + 2) \times x$ ✓

$\Rightarrow x(3x + 2) = 1$

$\Rightarrow 3x^2 + 2x = 1$

$\Rightarrow 3x^2 + 2x - 1 = 0$ ✓

$\Rightarrow (3x - 1)(x + 1) = 0$ ✓

$\Rightarrow 3x - 1 = 0$ or $x + 1 = 0$

$\Rightarrow 3x = 1$ or $x = -1$

$\Rightarrow x = \frac{1}{3}$ or $x = -1$

$x = -1$ is not a valid solution as lengths are positive

So $x = \frac{1}{3}$ is the only solution. ✓

5 marks

Area
- Obtaining $x(3x + 2)$ gains this mark.

Equation
- This mark is gained by setting up the equation: Area expression = 1.

Standard form
- Once a quadratic equation is recognised you should rearrange the terms into the standard form: $ax^2 + bx + c = 0$, i.e. x^2 term, followed by 'x' term then the constant.

Factorising
- Having factorised the quadratic expression you should multiply the brackets out to check that the factorisation is correct:

$$(3x - 1)(x + 1) = 3x^2 + 3x - x - 1 = 3x^2 + 2x - 1$$

Solution
- In many cases a mark will be assigned for rejecting an invalid solution like $x = -1$ and so this statement should be clear.

NOTES: 4·7 page 42

Practice Exam D: Paper 2 Worked Answers

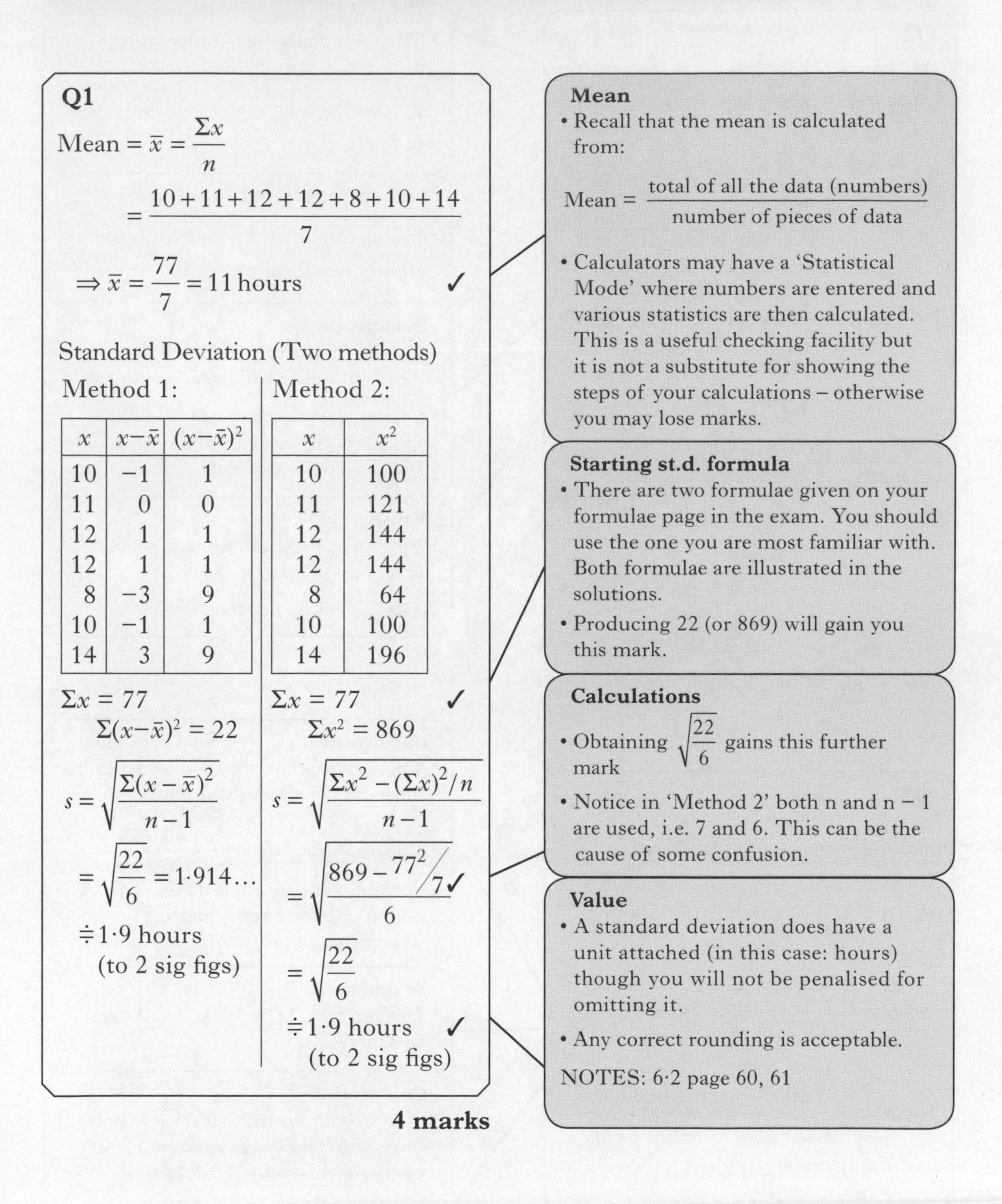

Q1

$$\text{Mean} = \bar{x} = \frac{\Sigma x}{n}$$

$$= \frac{10+11+12+12+8+10+14}{7}$$

$$\Rightarrow \bar{x} = \frac{77}{7} = 11 \text{ hours} \quad ✓$$

Standard Deviation (Two methods)

Method 1:

x	$x-\bar{x}$	$(x-\bar{x})^2$
10	−1	1
11	0	0
12	1	1
12	1	1
8	−3	9
10	−1	1
14	3	9

$\Sigma x = 77$

$\Sigma(x-\bar{x})^2 = 22$

$$s = \sqrt{\frac{\Sigma(x-\bar{x})^2}{n-1}}$$

$$= \sqrt{\frac{22}{6}} = 1{\cdot}914\ldots$$

$\doteqdot 1{\cdot}9$ hours (to 2 sig figs)

Method 2:

x	x^2
10	100
11	121
12	144
12	144
8	64
10	100
14	196

$\Sigma x = 77$ ✓

$\Sigma x^2 = 869$

$$s = \sqrt{\frac{\Sigma x^2 - (\Sigma x)^2/n}{n-1}}$$

$$= \sqrt{\frac{869 - 77^2/7}{6}} \quad ✓$$

$$= \sqrt{\frac{22}{6}}$$

$\doteqdot 1{\cdot}9$ hours ✓ (to 2 sig figs)

4 marks

Mean

- Recall that the mean is calculated from:

$$\text{Mean} = \frac{\text{total of all the data (numbers)}}{\text{number of pieces of data}}$$

- Calculators may have a 'Statistical Mode' where numbers are entered and various statistics are then calculated. This is a useful checking facility but it is not a substitute for showing the steps of your calculations – otherwise you may lose marks.

Starting st.d. formula

- There are two formulae given on your formulae page in the exam. You should use the one you are most familiar with. Both formulae are illustrated in the solutions.
- Producing 22 (or 869) will gain you this mark.

Calculations

- Obtaining $\sqrt{\frac{22}{6}}$ gains this further mark
- Notice in 'Method 2' both n and n − 1 are used, i.e. 7 and 6. This can be the cause of some confusion.

Value

- A standard deviation does have a unit attached (in this case: hours) though you will not be penalised for omitting it.
- Any correct rounding is acceptable.

NOTES: 6·2 page 60, 61

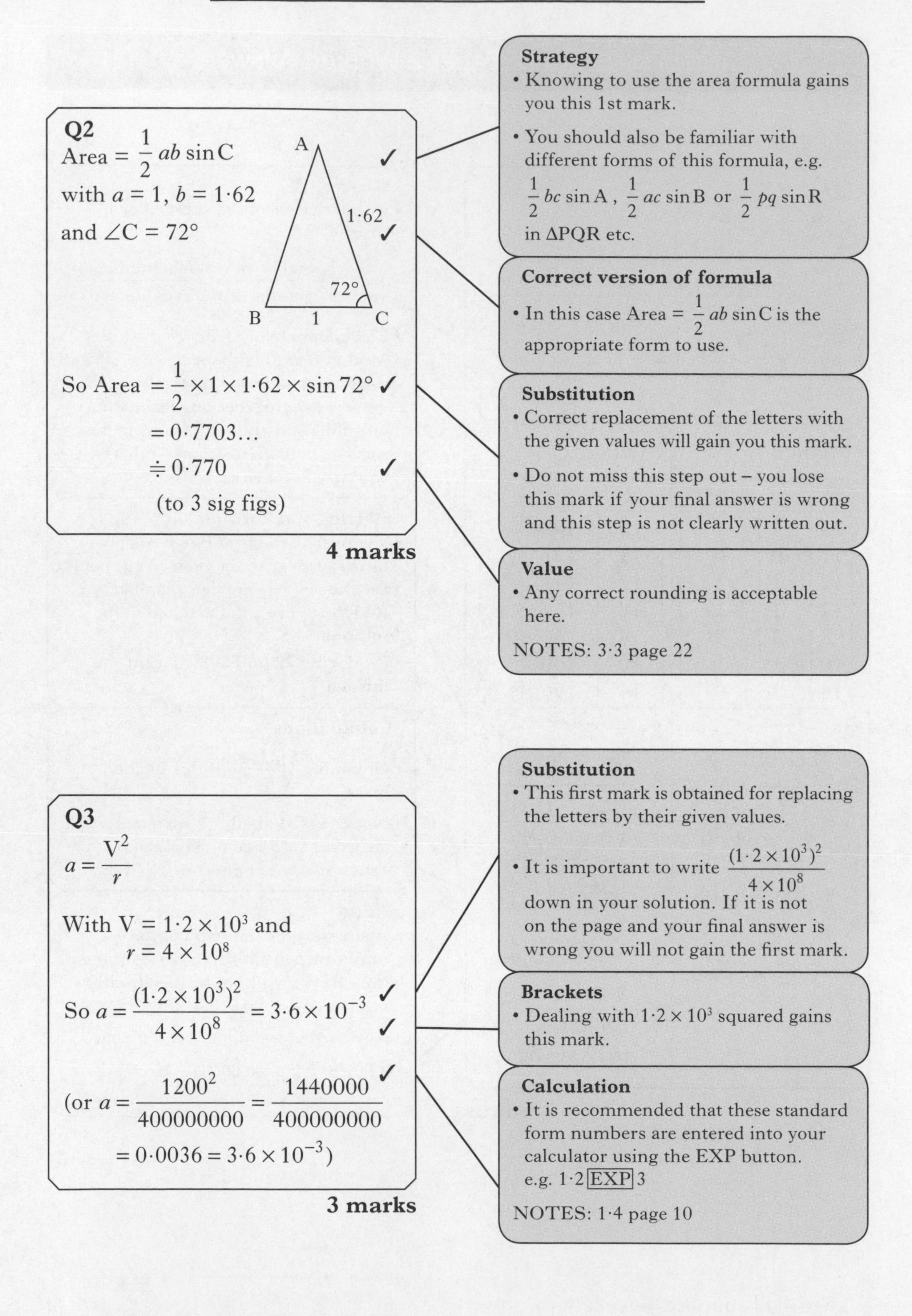

Q2

Area $= \frac{1}{2}ab\sin C$ ✓

with $a = 1$, $b = 1{\cdot}62$

and $\angle C = 72°$ ✓

So Area $= \frac{1}{2} \times 1 \times 1{\cdot}62 \times \sin 72°$ ✓

$= 0{\cdot}7703\ldots$

$\doteqdot 0{\cdot}770$ ✓

(to 3 sig figs)

4 marks

Strategy
- Knowing to use the area formula gains you this 1st mark.
- You should also be familiar with different forms of this formula, e.g. $\frac{1}{2}bc\sin A$, $\frac{1}{2}ac\sin B$ or $\frac{1}{2}pq\sin R$ in ΔPQR etc.

Correct version of formula
- In this case Area $= \frac{1}{2}ab\sin C$ is the appropriate form to use.

Substitution
- Correct replacement of the letters with the given values will gain you this mark.
- Do not miss this step out – you lose this mark if your final answer is wrong and this step is not clearly written out.

Value
- Any correct rounding is acceptable here.

NOTES: 3·3 page 22

Q3

$a = \frac{V^2}{r}$

With $V = 1{\cdot}2 \times 10^3$ and $r = 4 \times 10^8$

So $a = \frac{(1{\cdot}2 \times 10^3)^2}{4 \times 10^8} = 3{\cdot}6 \times 10^{-3}$ ✓ ✓ ✓

(or $a = \frac{1200^2}{400000000} = \frac{1440000}{400000000} = 0{\cdot}0036 = 3{\cdot}6 \times 10^{-3}$)

3 marks

Substitution
- This first mark is obtained for replacing the letters by their given values.
- It is important to write $\frac{(1{\cdot}2 \times 10^3)^2}{4 \times 10^8}$ down in your solution. If it is not on the page and your final answer is wrong you will not gain the first mark.

Brackets
- Dealing with $1{\cdot}2 \times 10^3$ squared gains this mark.

Calculation
- It is recommended that these standard form numbers are entered into your calculator using the EXP button. e.g. 1·2 EXP 3

NOTES: 1·4 page 10

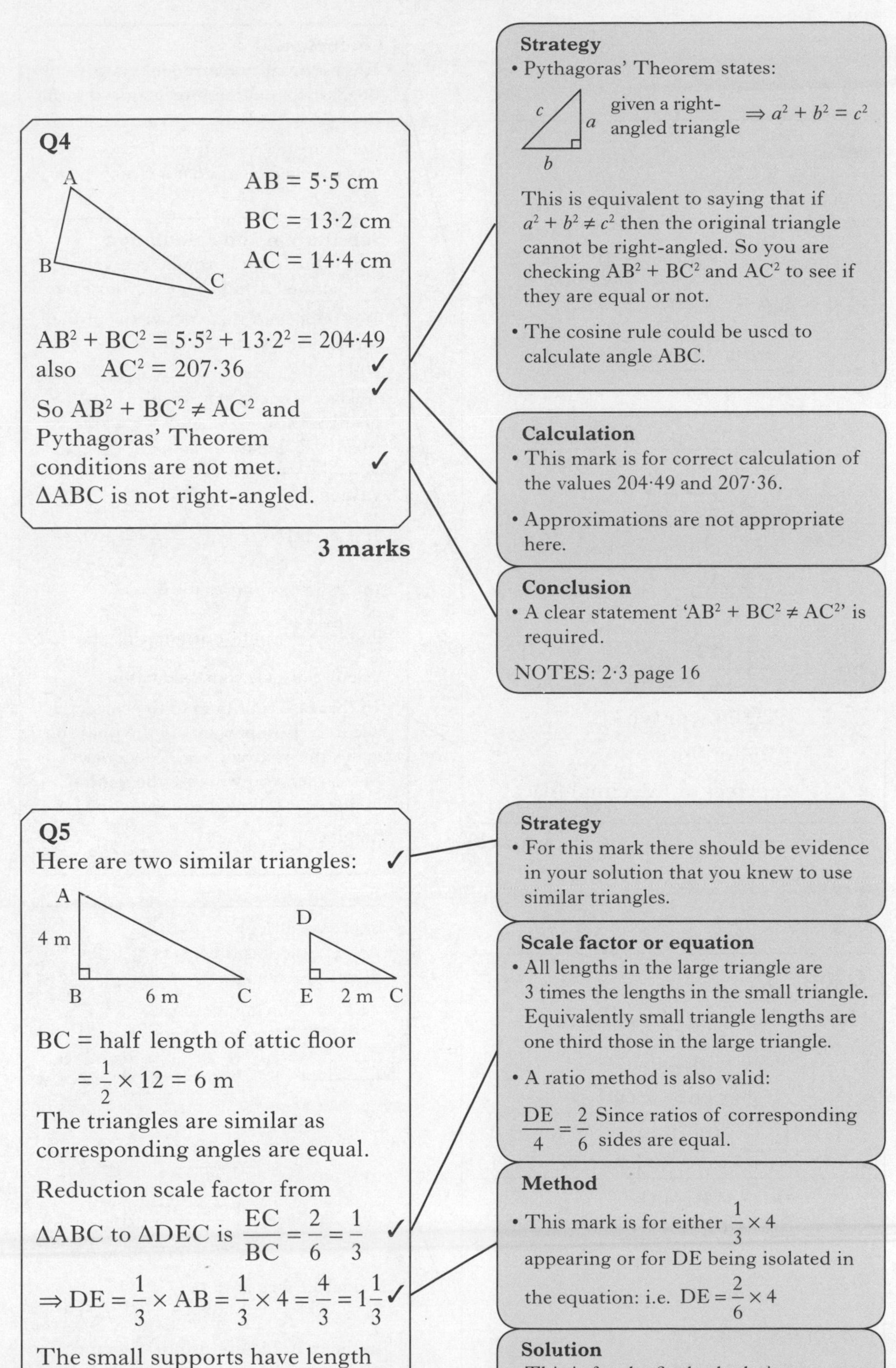

Q4

AB = 5·5 cm

BC = 13·2 cm

AC = 14·4 cm

$AB^2 + BC^2 = 5{\cdot}5^2 + 13{\cdot}2^2 = 204{\cdot}49$ ✓

also $AC^2 = 207{\cdot}36$ ✓

So $AB^2 + BC^2 \neq AC^2$ and Pythagoras' Theorem conditions are not met.
ΔABC is not right-angled. ✓

3 marks

Strategy

- Pythagoras' Theorem states: given a right-angled triangle $\Rightarrow a^2 + b^2 = c^2$

This is equivalent to saying that if $a^2 + b^2 \neq c^2$ then the original triangle cannot be right-angled. So you are checking $AB^2 + BC^2$ and AC^2 to see if they are equal or not.

- The cosine rule could be used to calculate angle ABC.

Calculation

- This mark is for correct calculation of the values 204·49 and 207·36.
- Approximations are not appropriate here.

Conclusion

- A clear statement '$AB^2 + BC^2 \neq AC^2$' is required.

NOTES: 2·3 page 16

Q5

Here are two similar triangles: ✓

BC = half length of attic floor

$= \frac{1}{2} \times 12 = 6$ m

The triangles are similar as corresponding angles are equal.

Reduction scale factor from ΔABC to ΔDEC is $\frac{EC}{BC} = \frac{2}{6} = \frac{1}{3}$ ✓

$\Rightarrow DE = \frac{1}{3} \times AB = \frac{1}{3} \times 4 = \frac{4}{3} = 1\frac{1}{3}$ ✓

The small supports have length $1\frac{1}{3}$ m ✓

4 marks

Strategy

- For this mark there should be evidence in your solution that you knew to use similar triangles.

Scale factor or equation

- All lengths in the large triangle are 3 times the lengths in the small triangle. Equivalently small triangle lengths are one third those in the large triangle.
- A ratio method is also valid:

$\frac{DE}{4} = \frac{2}{6}$ Since ratios of corresponding sides are equal.

Method

- This mark is for either $\frac{1}{3} \times 4$ appearing or for DE being isolated in the equation: i.e. $DE = \frac{2}{6} \times 4$

Solution

- This is for the final calculation.

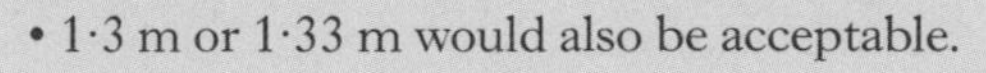

- 1·3 m or 1·33 m would also be acceptable.

NOTES: 2·2 page 15

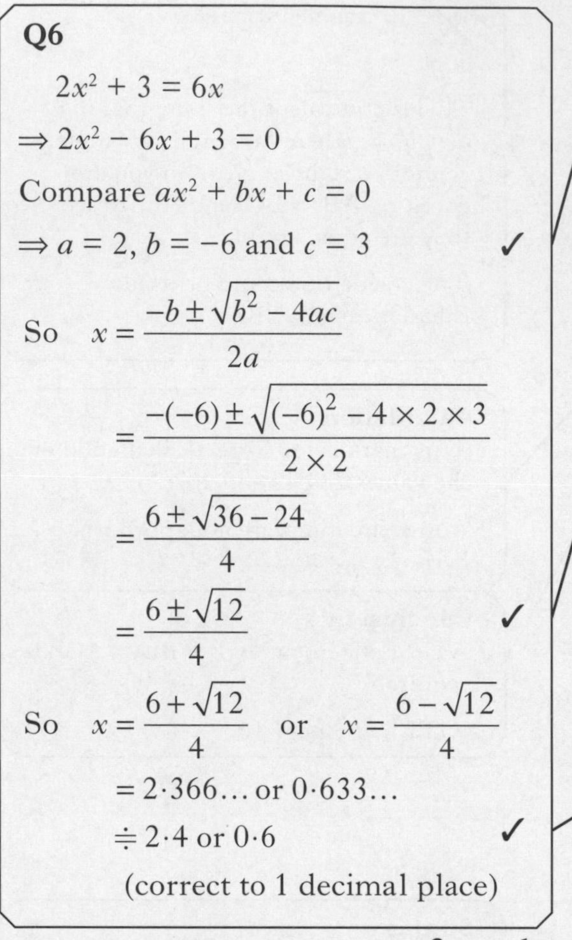

Q6

$$2x^2 + 3 = 6x$$

$\Rightarrow 2x^2 - 6x + 3 = 0$

Compare $ax^2 + bx + c = 0$

$\Rightarrow a = 2,\ b = -6$ and $c = 3$ ✓

So $x = \dfrac{-b \pm \sqrt{b^2 - 4ac}}{2a}$

$= \dfrac{-(-6) \pm \sqrt{(-6)^2 - 4 \times 2 \times 3}}{2 \times 2}$

$= \dfrac{6 \pm \sqrt{36 - 24}}{4}$

$= \dfrac{6 \pm \sqrt{12}}{4}$ ✓

So $x = \dfrac{6 + \sqrt{12}}{4}$ or $x = \dfrac{6 - \sqrt{12}}{4}$

$= 2{\cdot}366\ldots$ or $0{\cdot}633\ldots$

$\doteqdot 2{\cdot}4$ or $0{\cdot}6$ ✓

(correct to 1 decimal place)

3 marks

Coefficients

- It is essential to rearrange this quadratic equation into standard form: $ax^2 + bx + c = 0$
- Take care over negatives. The x-term $-6x$ has coefficient -6 not 6, so $b = -6$ not $b = 6$.

Substitution and calculation

- The 'Quadratic Formula' is given to you on the formula page in your exam.
- It is important that you write out the formula and state the values of a, b and c.
- Take care over negatives $(-6)^2 = 36$: positive. Also $-b$ when $b = -6$ gives $-(-6) = 6$: positive.

Values

- It is useful to rewrite $\dfrac{6 \pm \sqrt{12}}{4}$ giving the two possibilities i.e. $\dfrac{6 + \sqrt{12}}{4}$ and $\dfrac{6 - \sqrt{12}}{4}$ before attempting the calculations on your calculator.
- In this case rounding to the correct accuracy is important. If the question states the accuracy, e.g. 1 decimal place, then you will lose the mark if you do not follow this instruction.

NOTES: 4·7 page 43

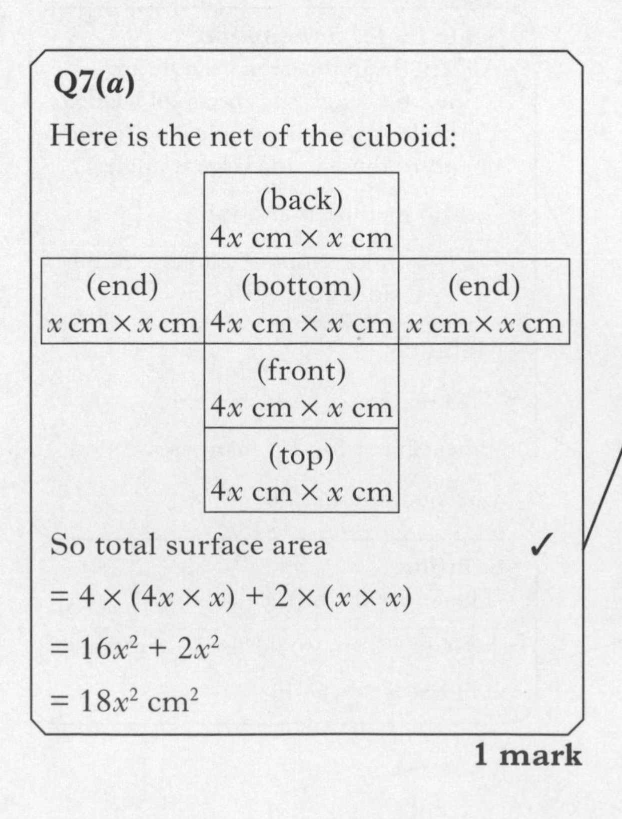

Q7(*a*)

Here is the net of the cuboid:

	(back) $4x$ cm $\times$ x cm	
(end) x cm $\times$ x cm	(bottom) $4x$ cm $\times$ x cm	(end) x cm $\times$ x cm
	(front) $4x$ cm $\times$ x cm	
	(top) $4x$ cm $\times$ x cm	

So total surface area ✓

$= 4 \times (4x \times x) + 2 \times (x \times x)$

$= 16x^2 + 2x^2$

$= 18x^2\ \text{cm}^2$

1 mark

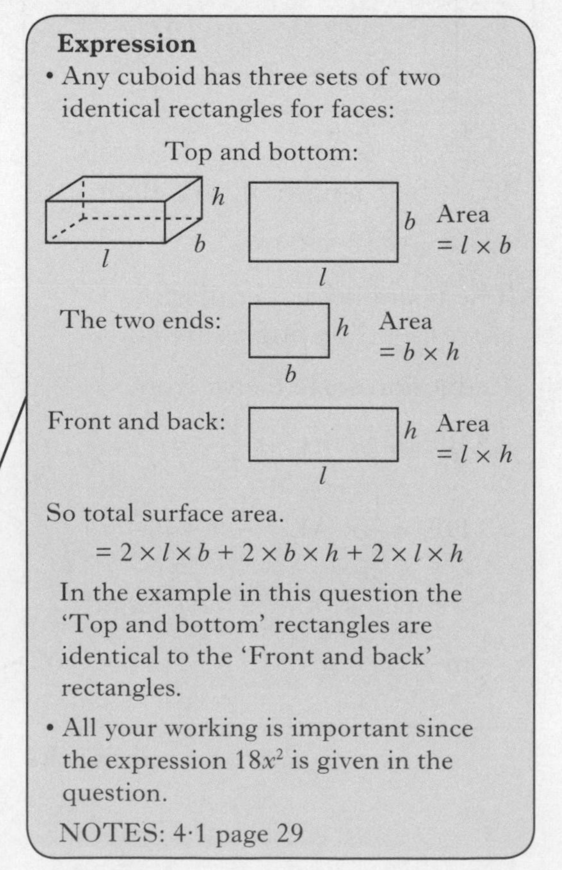

Expression

- Any cuboid has three sets of two identical rectangles for faces:

So total surface area.

$= 2 \times l \times b + 2 \times b \times h + 2 \times l \times h$

In the example in this question the 'Top and bottom' rectangles are identical to the 'Front and back' rectangles.

- All your working is important since the expression $18x^2$ is given in the question.

NOTES: 4·1 page 29

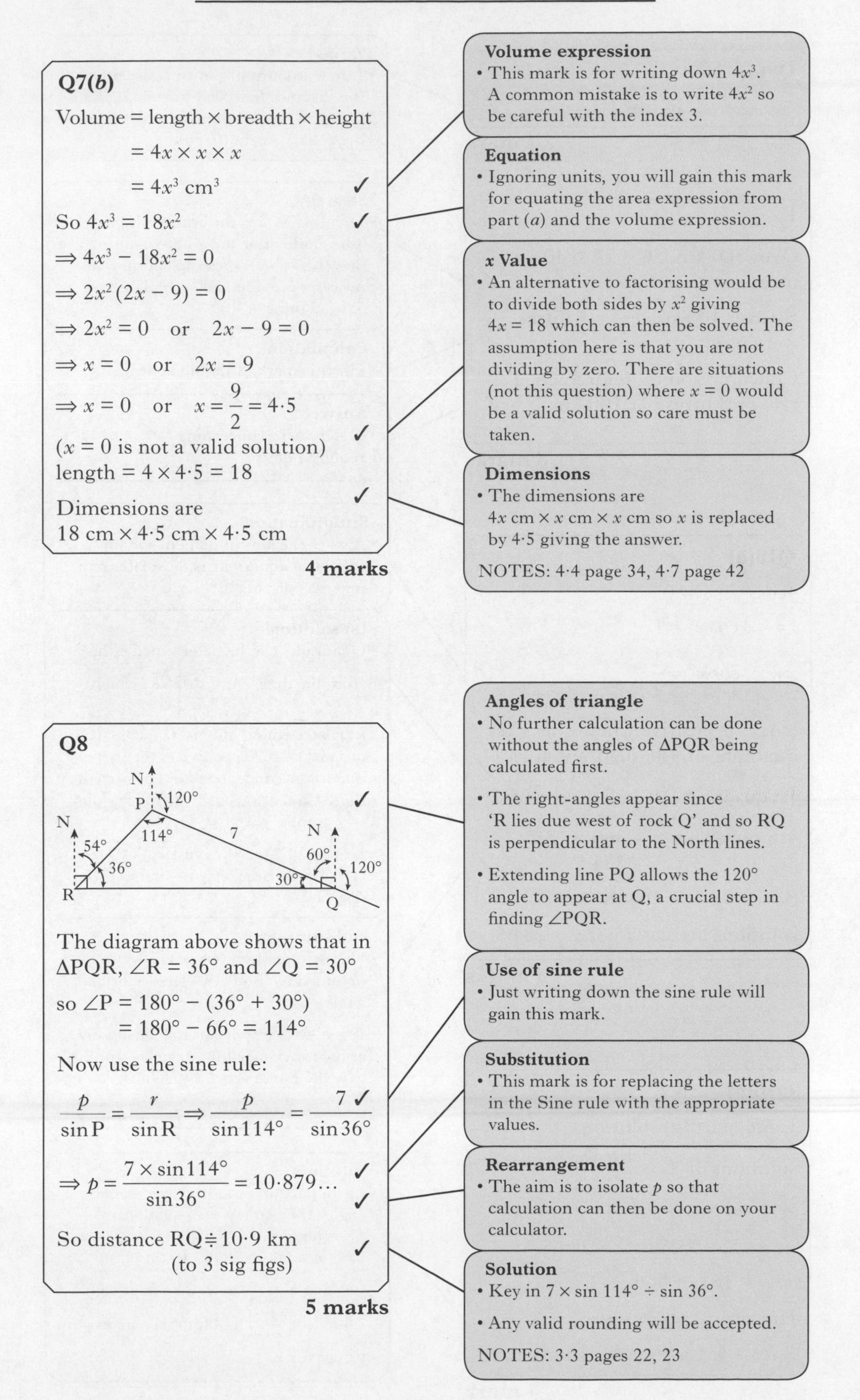

Q7(*b*)

Volume = length × breadth × height

$= 4x \times x \times x$

$= 4x^3\ \text{cm}^3$ ✓

So $4x^3 = 18x^2$ ✓

$\Rightarrow 4x^3 - 18x^2 = 0$

$\Rightarrow 2x^2(2x - 9) = 0$

$\Rightarrow 2x^2 = 0 \quad \text{or} \quad 2x - 9 = 0$

$\Rightarrow x = 0 \quad \text{or} \quad 2x = 9$

$\Rightarrow x = 0 \quad \text{or} \quad x = \frac{9}{2} = 4{\cdot}5$ ✓

($x = 0$ is not a valid solution)
length = $4 \times 4{\cdot}5 = 18$ ✓

Dimensions are
18 cm × 4·5 cm × 4·5 cm

4 marks

Volume expression
- This mark is for writing down $4x^3$. A common mistake is to write $4x^2$ so be careful with the index 3.

Equation
- Ignoring units, you will gain this mark for equating the area expression from part (*a*) and the volume expression.

***x* Value**
- An alternative to factorising would be to divide both sides by x^2 giving $4x = 18$ which can then be solved. The assumption here is that you are not dividing by zero. There are situations (not this question) where $x = 0$ would be a valid solution so care must be taken.

Dimensions
- The dimensions are $4x$ cm × x cm × x cm so x is replaced by 4·5 giving the answer.

NOTES: 4·4 page 34, 4·7 page 42

Q8

✓

The diagram above shows that in ΔPQR, $\angle R = 36°$ and $\angle Q = 30°$

so $\angle P = 180° - (36° + 30°)$
$= 180° - 66° = 114°$

Now use the sine rule:

$$\frac{p}{\sin P} = \frac{r}{\sin R} \Rightarrow \frac{p}{\sin 114°} = \frac{7}{\sin 36°}$$ ✓

$$\Rightarrow p = \frac{7 \times \sin 114°}{\sin 36°} = 10{\cdot}879\ldots$$ ✓ ✓

So distance RQ ≑ 10·9 km
(to 3 sig figs) ✓

5 marks

Angles of triangle
- No further calculation can be done without the angles of ΔPQR being calculated first.
- The right-angles appear since 'R lies due west of rock Q' and so RQ is perpendicular to the North lines.
- Extending line PQ allows the 120° angle to appear at Q, a crucial step in finding $\angle PQR$.

Use of sine rule
- Just writing down the sine rule will gain this mark.

Substitution
- This mark is for replacing the letters in the Sine rule with the appropriate values.

Rearrangement
- The aim is to isolate p so that calculation can then be done on your calculator.

Solution
- Key in 7 × sin 114° ÷ sin 36°.
- Any valid rounding will be accepted.

NOTES: 3·3 pages 22, 23

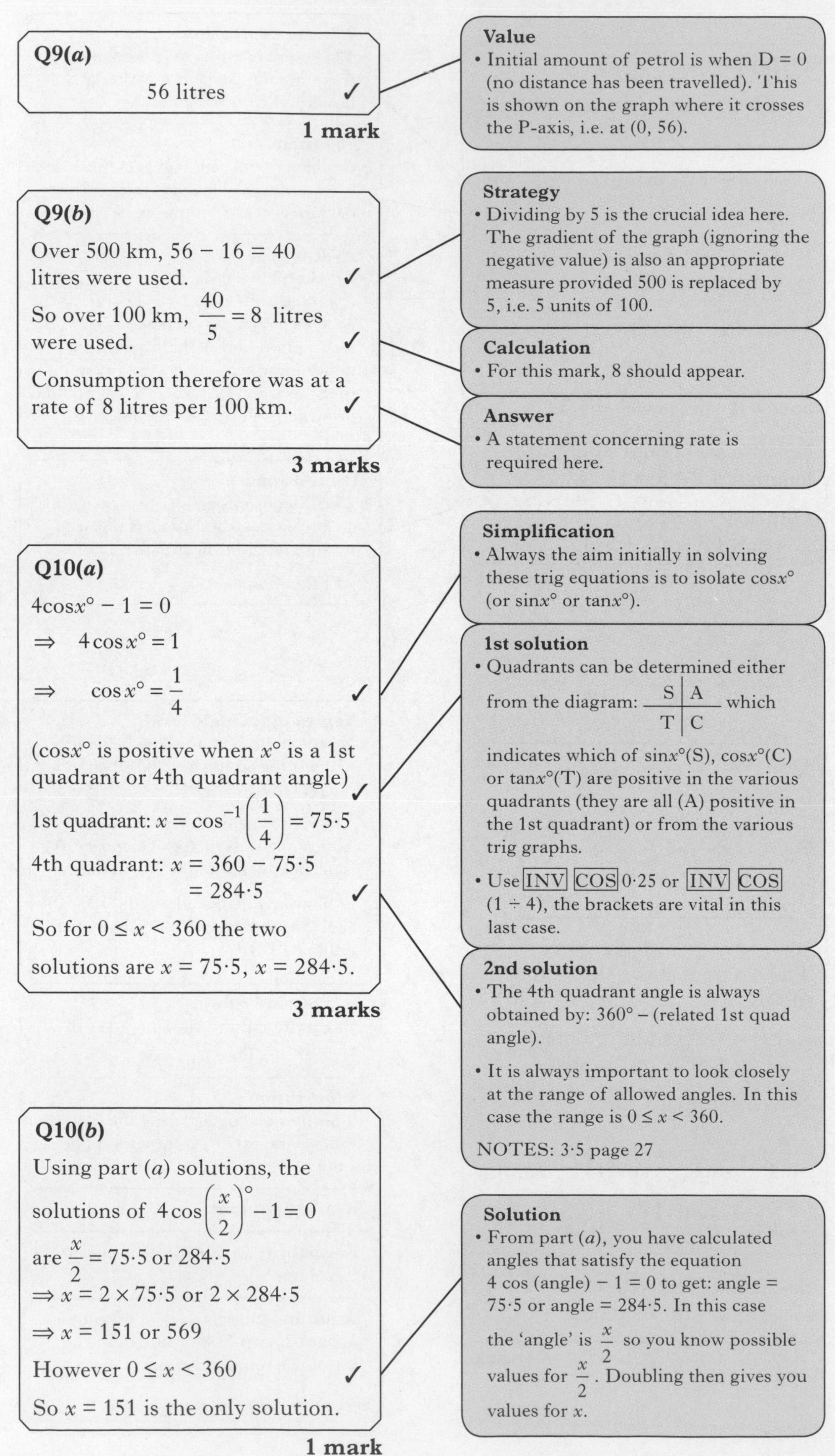

Q9(*a*)

56 litres ✓

1 mark

Value
- Initial amount of petrol is when D = 0 (no distance has been travelled). This is shown on the graph where it crosses the P-axis, i.e. at (0, 56).

Q9(*b*)

Over 500 km, 56 − 16 = 40 litres were used. ✓

So over 100 km, $\frac{40}{5} = 8$ litres were used. ✓

Consumption therefore was at a rate of 8 litres per 100 km. ✓

3 marks

Strategy
- Dividing by 5 is the crucial idea here. The gradient of the graph (ignoring the negative value) is also an appropriate measure provided 500 is replaced by 5, i.e. 5 units of 100.

Calculation
- For this mark, 8 should appear.

Answer
- A statement concerning rate is required here.

Q10(*a*)

$4\cos x° - 1 = 0$

$\Rightarrow \quad 4\cos x° = 1$

$\Rightarrow \quad \cos x° = \frac{1}{4}$ ✓

($\cos x°$ is positive when $x°$ is a 1st quadrant or 4th quadrant angle) ✓

1st quadrant: $x = \cos^{-1}\left(\frac{1}{4}\right) = 75{\cdot}5$

4th quadrant: $x = 360 - 75{\cdot}5$
$= 284{\cdot}5$ ✓

So for $0 \leq x < 360$ the two solutions are $x = 75{\cdot}5$, $x = 284{\cdot}5$.

3 marks

Simplification
- Always the aim initially in solving these trig equations is to isolate $\cos x°$ (or $\sin x°$ or $\tan x°$).

1st solution
- Quadrants can be determined either from the diagram:

S	A
T	C

which indicates which of $\sin x°$(S), $\cos x°$(C) or $\tan x°$(T) are positive in the various quadrants (they are all (A) positive in the 1st quadrant) or from the various trig graphs.
- Use [INV] [COS] 0·25 or [INV] [COS] (1 ÷ 4), the brackets are vital in this last case.

2nd solution
- The 4th quadrant angle is always obtained by: 360° – (related 1st quad angle).
- It is always important to look closely at the range of allowed angles. In this case the range is $0 \leq x < 360$.

NOTES: 3·5 page 27

Q10(*b*)

Using part (*a*) solutions, the solutions of $4\cos\left(\frac{x}{2}\right)^{\circ} - 1 = 0$

are $\frac{x}{2} = 75{\cdot}5$ or $284{\cdot}5$

$\Rightarrow x = 2 \times 75{\cdot}5$ or $2 \times 284{\cdot}5$

$\Rightarrow x = 151$ or 569

However $0 \leq x < 360$ ✓

So $x = 151$ is the only solution.

1 mark

Solution
- From part (*a*), you have calculated angles that satisfy the equation 4 cos (angle) − 1 = 0 to get: angle = 75·5 or angle = 284·5. In this case the 'angle' is $\frac{x}{2}$ so you know possible values for $\frac{x}{2}$. Doubling then gives you values for x.

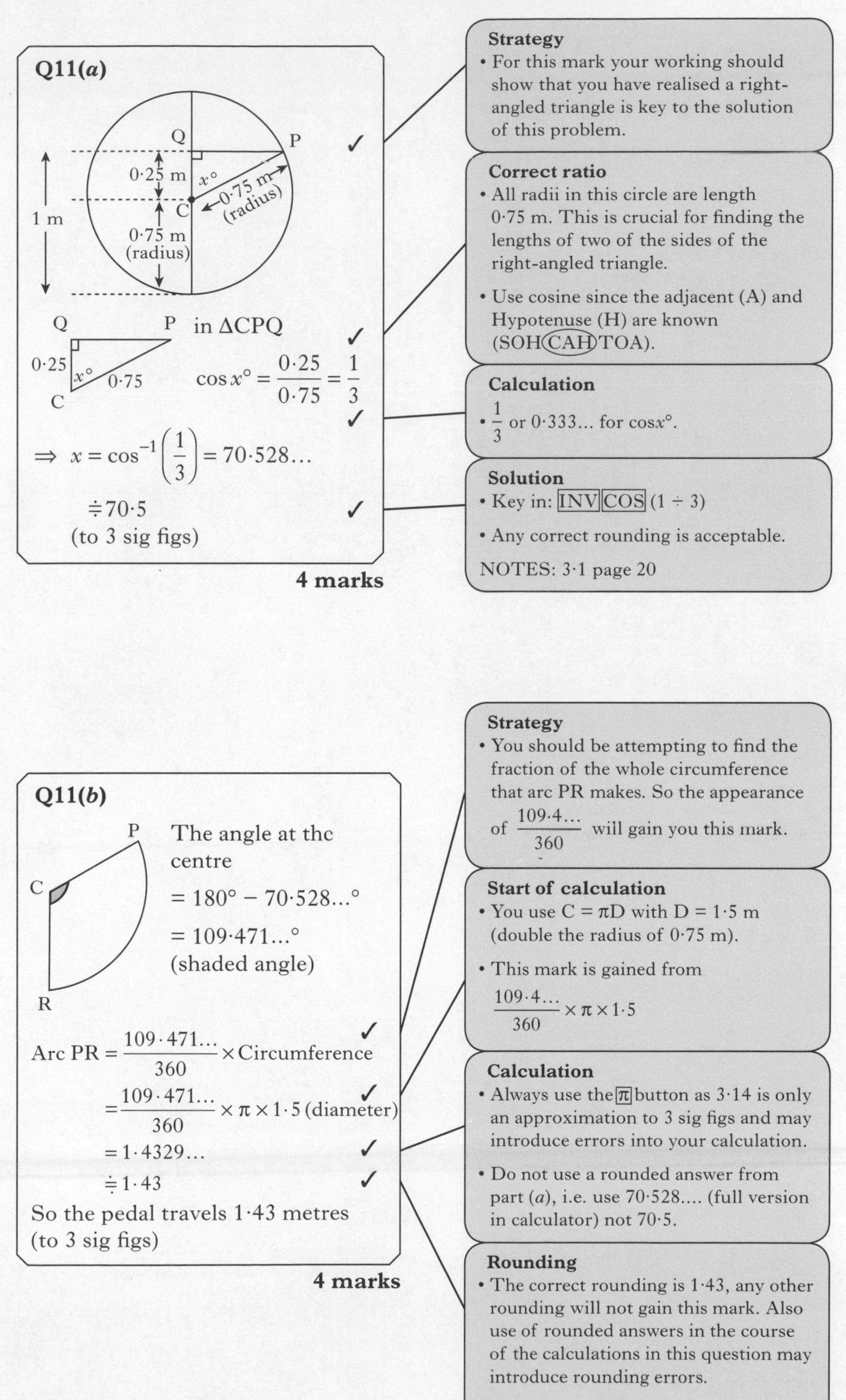

Q11(*a*)

in ΔCPQ ✓

$$\cos x° = \frac{0{\cdot}25}{0{\cdot}75} = \frac{1}{3}$$ ✓

$$\Rightarrow \; x = \cos^{-1}\left(\frac{1}{3}\right) = 70{\cdot}528\ldots$$ ✓

$\doteqdot 70{\cdot}5$
(to 3 sig figs) ✓

4 marks

Strategy
- For this mark your working should show that you have realised a right-angled triangle is key to the solution of this problem.

Correct ratio
- All radii in this circle are length 0·75 m. This is crucial for finding the lengths of two of the sides of the right-angled triangle.
- Use cosine since the adjacent (A) and Hypotenuse (H) are known (SOH(CAH)TOA).

Calculation
- $\frac{1}{3}$ or 0·333... for $\cos x°$.

Solution
- Key in: [INV][COS] (1 ÷ 3)
- Any correct rounding is acceptable.

NOTES: 3·1 page 20

Q11(*b*)

The angle at thc centre
$= 180° - 70{\cdot}528\ldots°$
$= 109{\cdot}471\ldots°$
(shaded angle)

$$\text{Arc PR} = \frac{109{\cdot}471\ldots}{360} \times \text{Circumference}$$ ✓

$$= \frac{109{\cdot}471\ldots}{360} \times \pi \times 1{\cdot}5 \text{ (diameter)}$$ ✓

$= 1{\cdot}4329\ldots$ ✓

$\doteqdot 1{\cdot}43$ ✓

So the pedal travels 1·43 metres (to 3 sig figs)

4 marks

Strategy
- You should be attempting to find the fraction of the whole circumference that arc PR makes. So the appearance of $\frac{109{\cdot}4\ldots}{360}$ will gain you this mark.

Start of calculation
- You use $C = \pi D$ with $D = 1{\cdot}5$ m (double the radius of 0·75 m).
- This mark is gained from $\frac{109{\cdot}4\ldots}{360} \times \pi \times 1{\cdot}5$

Calculation
- Always use the [π] button as 3·14 is only an approximation to 3 sig figs and may introduce errors into your calculation.
- Do not use a rounded answer from part (*a*), i.e. use 70·528.... (full version in calculator) not 70·5.

Rounding
- The correct rounding is 1·43, any other rounding will not gain this mark. Also use of rounded answers in the course of the calculations in this question may introduce rounding errors.

NOTES: 2·4 page 17

Practice Exam E: Paper 1 Worked Answers

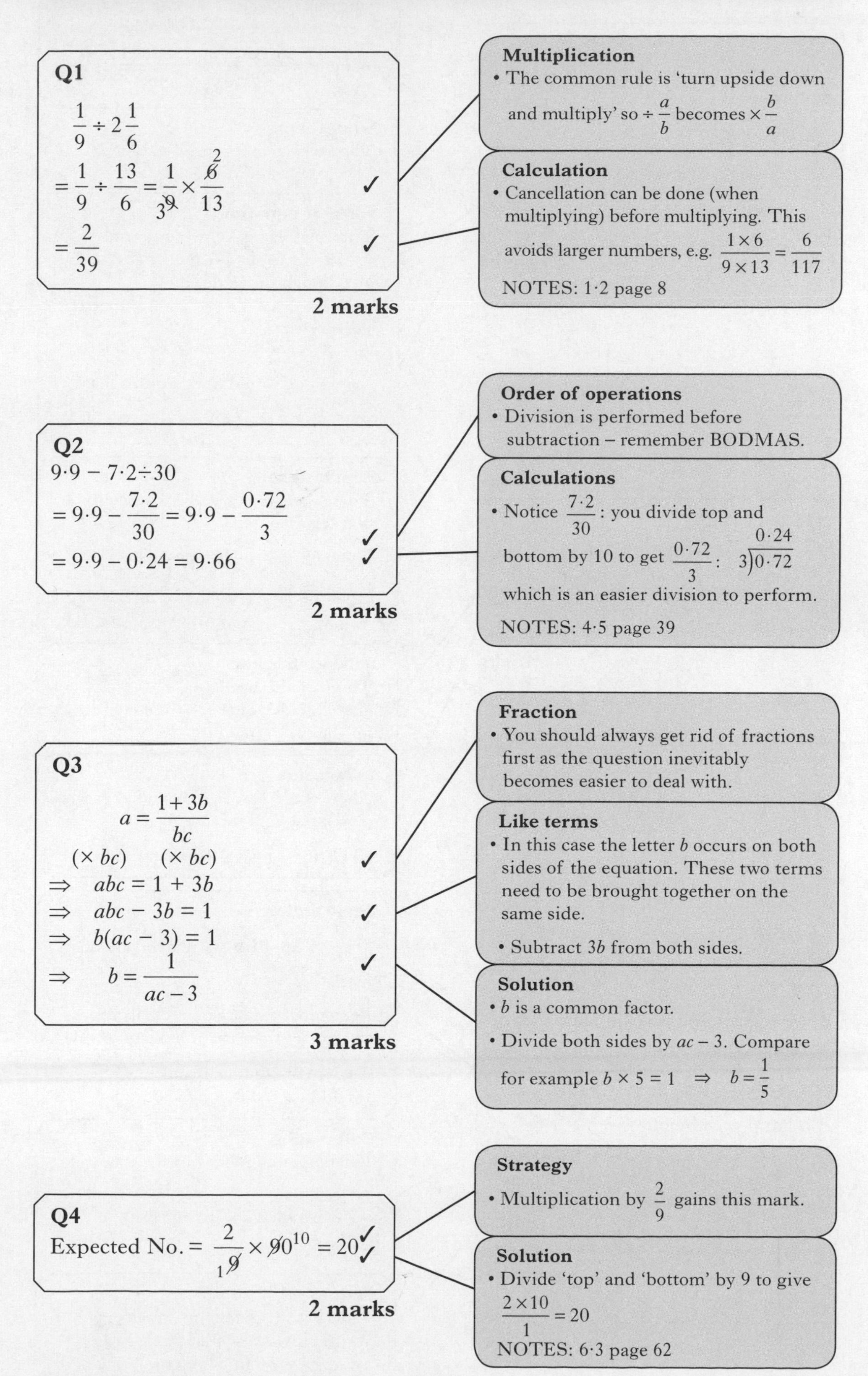

Q1

$$\frac{1}{9} \div 2\frac{1}{6}$$

$$= \frac{1}{9} \div \frac{13}{6} = \frac{1}{\not{9}_3} \times \frac{\not{6}^2}{13}$$ ✓

$$= \frac{2}{39}$$ ✓

2 marks

Multiplication
- The common rule is 'turn upside down and multiply' so $\div \frac{a}{b}$ becomes $\times \frac{b}{a}$

Calculation
- Cancellation can be done (when multiplying) before multiplying. This avoids larger numbers, e.g. $\frac{1 \times 6}{9 \times 13} = \frac{6}{117}$

NOTES: 1·2 page 8

Q2

$9{\cdot}9 - 7{\cdot}2 \div 30$

$$= 9{\cdot}9 - \frac{7{\cdot}2}{30} = 9{\cdot}9 - \frac{0{\cdot}72}{3}$$ ✓

$$= 9{\cdot}9 - 0{\cdot}24 = 9{\cdot}66$$ ✓

2 marks

Order of operations
- Division is performed before subtraction – remember BODMAS.

Calculations
- Notice $\frac{7{\cdot}2}{30}$: you divide top and bottom by 10 to get $\frac{0{\cdot}72}{3}$: $3\overline{)0{\cdot}72}$ giving $0{\cdot}24$, which is an easier division to perform.

NOTES: 4·5 page 39

Q3

$$a = \frac{1+3b}{bc}$$

$(\times bc) \quad (\times bc)$ ✓

$\Rightarrow \quad abc = 1 + 3b$

$\Rightarrow \quad abc - 3b = 1$ ✓

$\Rightarrow \quad b(ac - 3) = 1$

$\Rightarrow \quad b = \frac{1}{ac-3}$ ✓

3 marks

Fraction
- You should always get rid of fractions first as the question inevitably becomes easier to deal with.

Like terms
- In this case the letter b occurs on both sides of the equation. These two terms need to be brought together on the same side.
- Subtract $3b$ from both sides.

Solution
- b is a common factor.
- Divide both sides by $ac - 3$. Compare for example $b \times 5 = 1 \Rightarrow b = \frac{1}{5}$

Q4

Expected No. $= \frac{2}{\not{9}_1} \times \not{90}^{10} = 20$ ✓✓

2 marks

Strategy
- Multiplication by $\frac{2}{9}$ gains this mark.

Solution
- Divide 'top' and 'bottom' by 9 to give $\frac{2 \times 10}{1} = 20$

NOTES: 6·3 page 62

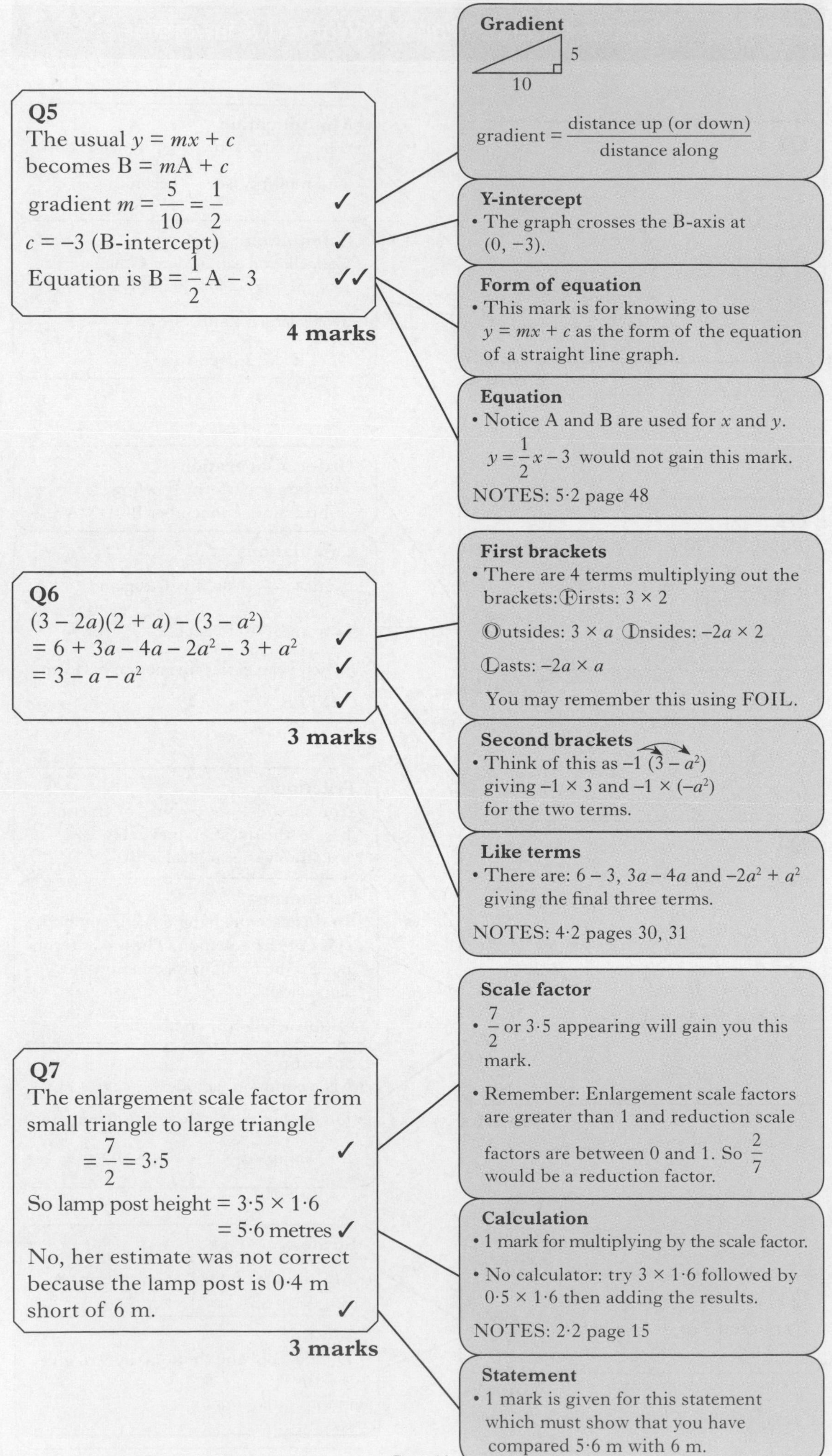

Q5

The usual $y = mx + c$

becomes $B = mA + c$

gradient $m = \frac{5}{10} = \frac{1}{2}$ ✓

$c = -3$ (B-intercept) ✓

Equation is $B = \frac{1}{2}A - 3$ ✓✓

4 marks

Gradient

(triangle: 10 along, 5 up)

gradient $= \frac{\text{distance up (or down)}}{\text{distance along}}$

Y-intercept

- The graph crosses the B-axis at (0, −3).

Form of equation

- This mark is for knowing to use $y = mx + c$ as the form of the equation of a straight line graph.

Equation

- Notice A and B are used for x and y.

  $y = \frac{1}{2}x - 3$ would not gain this mark.

NOTES: 5·2 page 48

Q6

$(3 - 2a)(2 + a) - (3 - a^2)$ ✓

$= 6 + 3a - 4a - 2a^2 - 3 + a^2$ ✓

$= 3 - a - a^2$ ✓

3 marks

First brackets

- There are 4 terms multiplying out the brackets: Ⓕirsts: 3×2

  Ⓞutsides: $3 \times a$ Ⓘnsides: $-2a \times 2$

  Ⓛasts: $-2a \times a$

  You may remember this using FOIL.

Second brackets

- Think of this as $-1\,(3 - a^2)$ giving -1×3 and $-1 \times (-a^2)$ for the two terms.

Like terms

- There are: $6 - 3$, $3a - 4a$ and $-2a^2 + a^2$ giving the final three terms.

NOTES: 4·2 pages 30, 31

Q7

The enlargement scale factor from small triangle to large triangle

$= \frac{7}{2} = 3{\cdot}5$ ✓

So lamp post height $= 3{\cdot}5 \times 1{\cdot}6$

$= 5{\cdot}6$ metres ✓

No, her estimate was not correct because the lamp post is 0·4 m short of 6 m. ✓

3 marks

Scale factor

- $\frac{7}{2}$ or 3·5 appearing will gain you this mark.
- Remember: Enlargement scale factors are greater than 1 and reduction scale factors are between 0 and 1. So $\frac{2}{7}$ would be a reduction factor.

Calculation

- 1 mark for multiplying by the scale factor.
- No calculator: try $3 \times 1{\cdot}6$ followed by $0{\cdot}5 \times 1{\cdot}6$ then adding the results.

NOTES: 2·2 page 15

Statement

- 1 mark is given for this statement which must show that you have compared 5·6 m with 6 m.

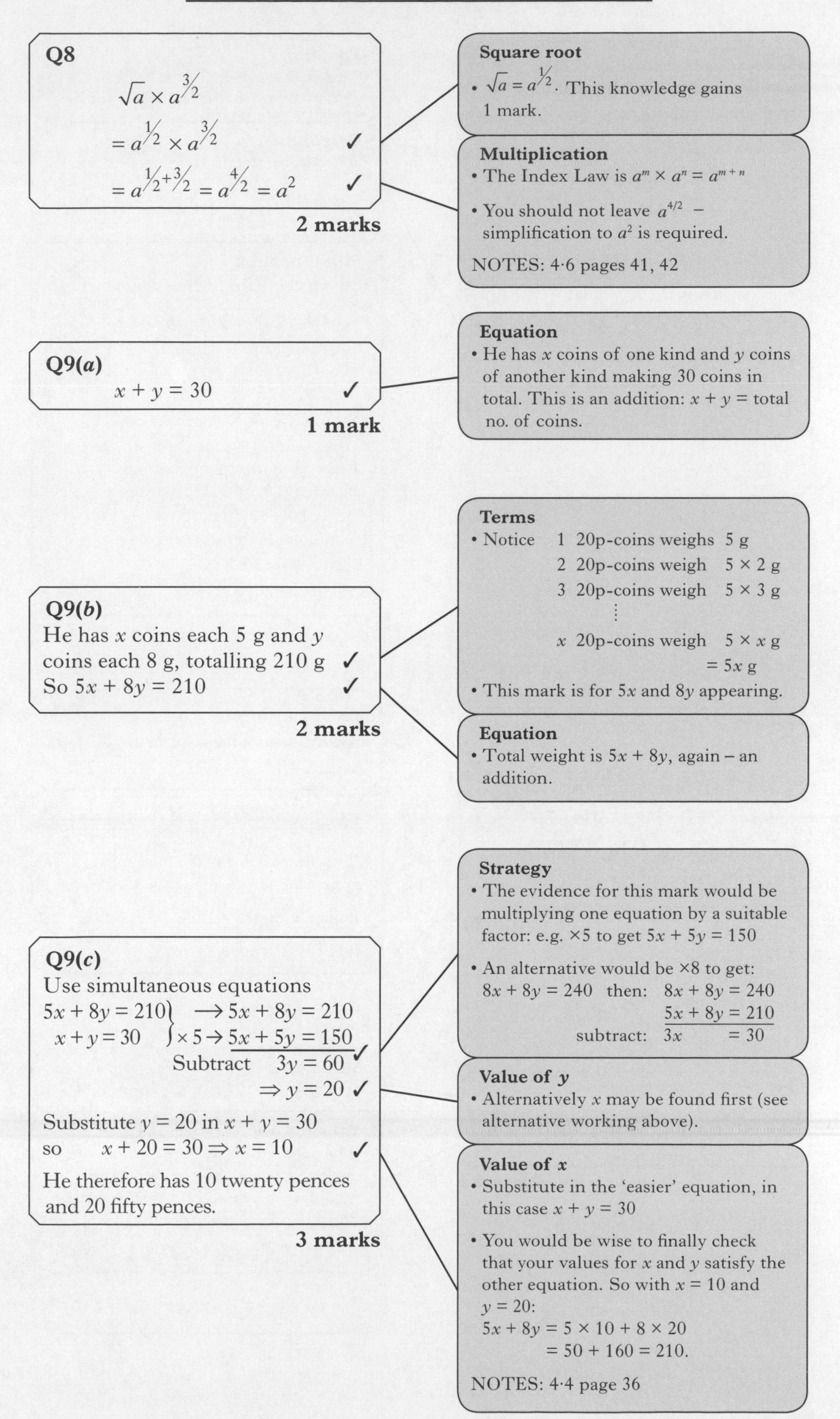

Q8

$$\sqrt{a} \times a^{3/2}$$
$$= a^{1/2} \times a^{3/2} \quad ✓$$
$$= a^{1/2 + 3/2} = a^{4/2} = a^2 \quad ✓$$

2 marks

Square root
- $\sqrt{a} = a^{1/2}$. This knowledge gains 1 mark.

Multiplication
- The Index Law is $a^m \times a^n = a^{m+n}$
- You should not leave $a^{4/2}$ – simplification to a^2 is required.

NOTES: 4·6 pages 41, 42

Q9(*a*)

$x + y = 30$ ✓

1 mark

Equation
- He has x coins of one kind and y coins of another kind making 30 coins in total. This is an addition: $x + y =$ total no. of coins.

Q9(*b*)

He has x coins each 5 g and y coins each 8 g, totalling 210 g ✓
So $5x + 8y = 210$ ✓

2 marks

Terms
- Notice 1 20p-coins weighs 5 g
  2 20p-coins weigh 5×2 g
  3 20p-coins weigh 5×3 g
  ⋮
  x 20p-coins weigh $5 \times x$ g $= 5x$ g
- This mark is for $5x$ and $8y$ appearing.

Equation
- Total weight is $5x + 8y$, again – an addition.

Q9(*c*)

Use simultaneous equations

$$\begin{aligned} 5x + 8y = 210 \quad &\longrightarrow 5x + 8y = 210 \\ x + y = 30 \quad \times 5 &\rightarrow 5x + 5y = 150 \quad ✓ \\ \text{Subtract} \quad & 3y = 60 \\ & \Rightarrow y = 20 \quad ✓ \end{aligned}$$

Substitute $y = 20$ in $x + y = 30$
so $x + 20 = 30 \Rightarrow x = 10$ ✓

He therefore has 10 twenty pences and 20 fifty pences.

3 marks

Strategy
- The evidence for this mark would be multiplying one equation by a suitable factor: e.g. ×5 to get $5x + 5y = 150$
- An alternative would be ×8 to get: $8x + 8y = 240$ then:

$$\begin{aligned} & 8x + 8y = 240 \\ & 5x + 8y = 210 \\ \text{subtract:} \quad & 3x \qquad = 30 \end{aligned}$$

Value of *y*
- Alternatively x may be found first (see alternative working above).

Value of *x*
- Substitute in the 'easier' equation, in this case $x + y = 30$
- You would be wise to finally check that your values for x and y satisfy the other equation. So with $x = 10$ and $y = 20$:

$$5x + 8y = 5 \times 10 + 8 \times 20$$
$$= 50 + 160 = 210.$$

NOTES: 4·4 page 36

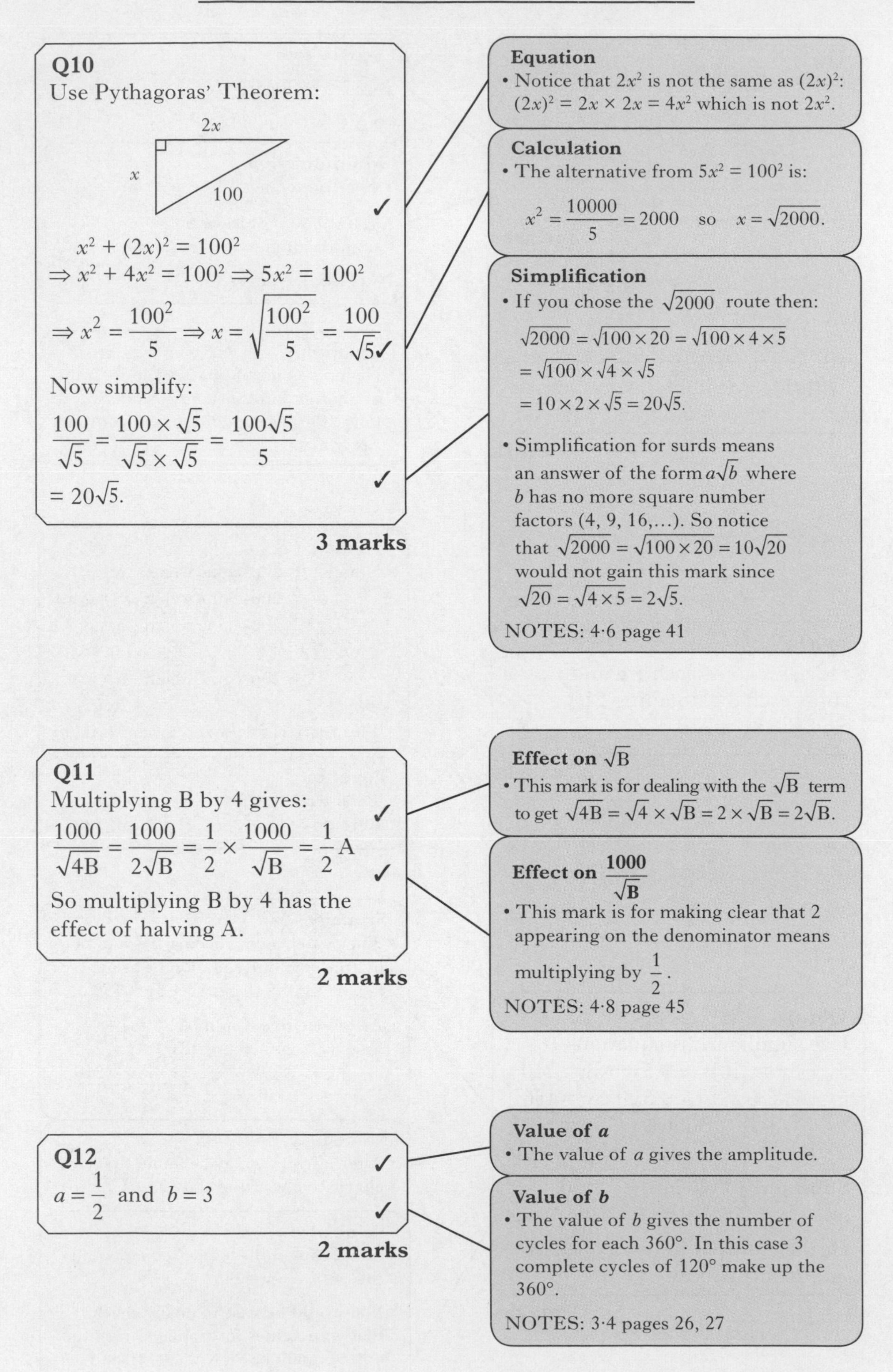

Q10

Use Pythagoras' Theorem:

$x^2 + (2x)^2 = 100^2$ ✓

$\Rightarrow x^2 + 4x^2 = 100^2 \Rightarrow 5x^2 = 100^2$

$\Rightarrow x^2 = \frac{100^2}{5} \Rightarrow x = \sqrt{\frac{100^2}{5}} = \frac{100}{\sqrt{5}}$ ✓

Now simplify:

$\frac{100}{\sqrt{5}} = \frac{100 \times \sqrt{5}}{\sqrt{5} \times \sqrt{5}} = \frac{100\sqrt{5}}{5}$

$= 20\sqrt{5}$. ✓

3 marks

Equation

- Notice that $2x^2$ is not the same as $(2x)^2$: $(2x)^2 = 2x \times 2x = 4x^2$ which is not $2x^2$.

Calculation

- The alternative from $5x^2 = 100^2$ is:

$$x^2 = \frac{10000}{5} = 2000 \quad \text{so} \quad x = \sqrt{2000}.$$

Simplification

- If you chose the $\sqrt{2000}$ route then:

$$\sqrt{2000} = \sqrt{100 \times 20} = \sqrt{100 \times 4 \times 5}$$
$$= \sqrt{100} \times \sqrt{4} \times \sqrt{5}$$
$$= 10 \times 2 \times \sqrt{5} = 20\sqrt{5}.$$

- Simplification for surds means an answer of the form $a\sqrt{b}$ where b has no more square number factors (4, 9, 16,…). So notice that $\sqrt{2000} = \sqrt{100 \times 20} = 10\sqrt{20}$ would not gain this mark since $\sqrt{20} = \sqrt{4 \times 5} = 2\sqrt{5}$.

NOTES: 4·6 page 41

Q11

Multiplying B by 4 gives: ✓

$\frac{1000}{\sqrt{4B}} = \frac{1000}{2\sqrt{B}} = \frac{1}{2} \times \frac{1000}{\sqrt{B}} = \frac{1}{2}A$ ✓

So multiplying B by 4 has the effect of halving A.

2 marks

Effect on $\sqrt{B}$

- This mark is for dealing with the $\sqrt{B}$ term to get $\sqrt{4B} = \sqrt{4} \times \sqrt{B} = 2 \times \sqrt{B} = 2\sqrt{B}$.

Effect on $\frac{1000}{\sqrt{B}}$

- This mark is for making clear that 2 appearing on the denominator means multiplying by $\frac{1}{2}$.

NOTES: 4·8 page 45

Q12

$a = \frac{1}{2}$ ✓ and $b = 3$ ✓

2 marks

Value of *a*

- The value of a gives the amplitude.

Value of *b*

- The value of b gives the number of cycles for each 360°. In this case 3 complete cycles of 120° make up the 360°.

NOTES: 3·4 pages 26, 27

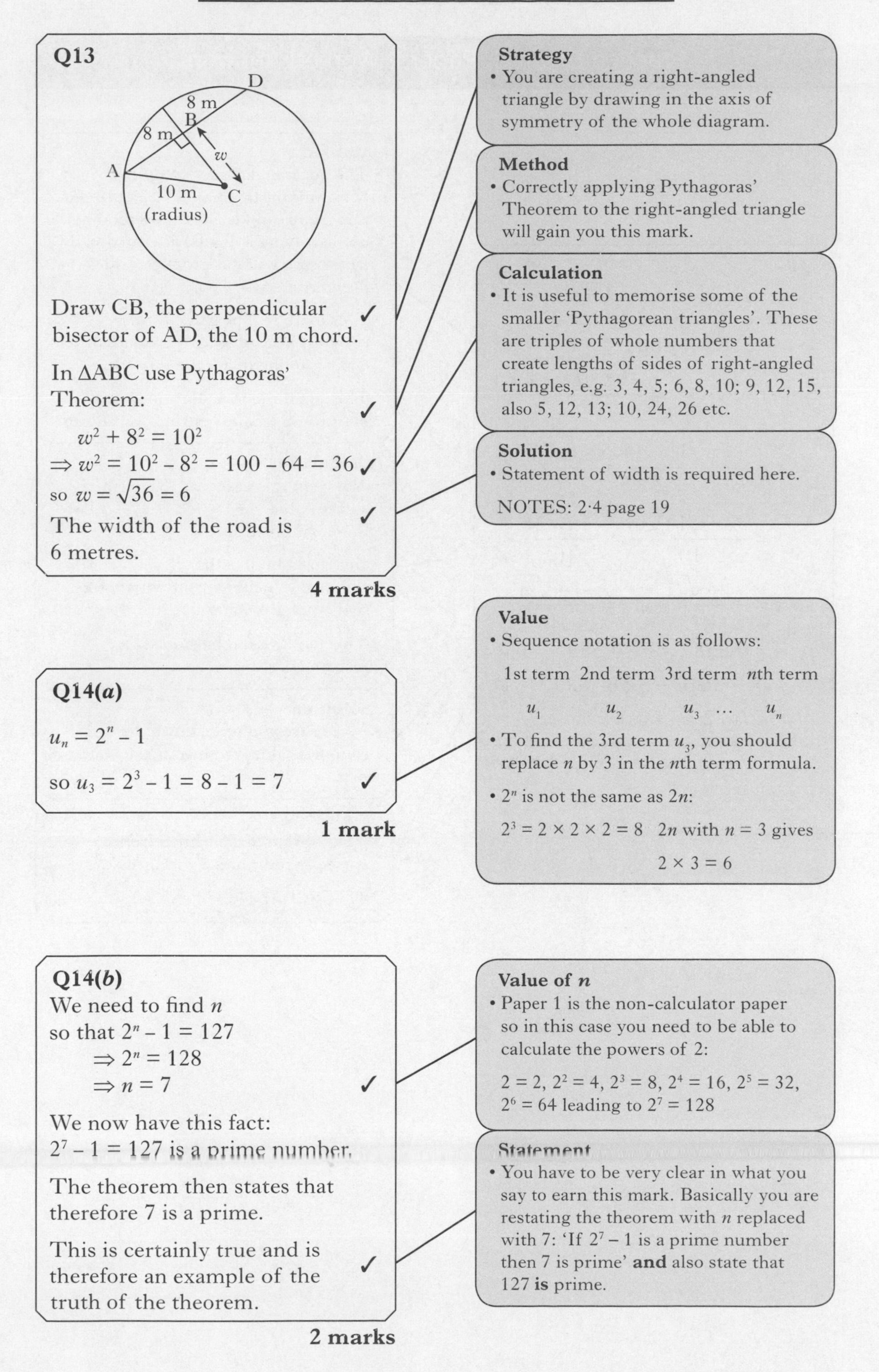

Q13

Draw CB, the perpendicular bisector of AD, the 10 m chord. ✓

In ΔABC use Pythagoras' Theorem: ✓

$w^2 + 8^2 = 10^2$

$\Rightarrow w^2 = 10^2 - 8^2 = 100 - 64 = 36$ ✓

so $w = \sqrt{36} = 6$

The width of the road is 6 metres. ✓

4 marks

Strategy
- You are creating a right-angled triangle by drawing in the axis of symmetry of the whole diagram.

Method
- Correctly applying Pythagoras' Theorem to the right-angled triangle will gain you this mark.

Calculation
- It is useful to memorise some of the smaller 'Pythagorean triangles'. These are triples of whole numbers that create lengths of sides of right-angled triangles, e.g. 3, 4, 5; 6, 8, 10; 9, 12, 15, also 5, 12, 13; 10, 24, 26 etc.

Solution
- Statement of width is required here.

NOTES: 2·4 page 19

Q14(*a*)

$u_n = 2^n - 1$

so $u_3 = 2^3 - 1 = 8 - 1 = 7$ ✓

1 mark

Value
- Sequence notation is as follows:

1st term	2nd term	3rd term		*n*th term
u_1	u_2	u_3	...	u_n

- To find the 3rd term u_3, you should replace n by 3 in the nth term formula.
- 2^n is not the same as $2n$:

$2^3 = 2 \times 2 \times 2 = 8$ $2n$ with $n = 3$ gives $2 \times 3 = 6$

Q14(*b*)

We need to find n so that $2^n - 1 = 127$

$\Rightarrow 2^n = 128$

$\Rightarrow n = 7$ ✓

We now have this fact: $2^7 - 1 = 127$ is a prime number.

The theorem then states that therefore 7 is a prime.

This is certainly true and is therefore an example of the truth of the theorem. ✓

2 marks

Value of n
- Paper 1 is the non-calculator paper so in this case you need to be able to calculate the powers of 2:

$2 = 2$, $2^2 = 4$, $2^3 = 8$, $2^4 = 16$, $2^5 = 32$, $2^6 = 64$ leading to $2^7 = 128$

Statement
- You have to be very clear in what you say to earn this mark. Basically you are restating the theorem with n replaced with 7: 'If $2^7 - 1$ is a prime number then 7 is prime' **and** also state that 127 **is** prime.

Practice Exam E: Paper 2 Worked Answers

Q1

$$5x^2 + x - 1 = 0$$

Compare $ax^2 + bx + c = 0$

$\Rightarrow a = 5, b = 1$ and $c = -1$ ✓

Substitute in $x = \dfrac{-b \pm \sqrt{b^2 - 4ac}}{2a}$

$$\Rightarrow x = \frac{-1 \pm \sqrt{1^2 - 4 \times 5 \times (-1)}}{2 \times 5}$$

$$= \frac{-1 \pm \sqrt{1+20}}{10} = \frac{-1 \pm \sqrt{21}}{10} \quad ✓$$

$$\text{so } x = \frac{-1 + \sqrt{21}}{10} \text{ or } x = \frac{-1 - \sqrt{21}}{10}$$

$$= 0{\cdot}358\ldots \text{ or } -0{\cdot}558\ldots \quad ✓$$

$$\doteqdot 0{\cdot}36 \text{ or } -0{\cdot}56 \quad ✓$$

(to 2 sig figs)

4 marks

Method

- The question asks for rounding (2 significant figures). This is the clue that factorising is not the method to be used. Solving a quadratic equation by factorising will not produce values that need rounding.
- This mark will be gained from correctly substituting the values of a, b and c into the quadratic formula.
- The quadratic formula will appear on your formula sheet during your exam.

Calculation

- Notice in this case that c is negative and so $-4ac = -4 \times 5 \times (-1) = 20$ is therefore positive.
- In a question like this, if you get a negative value under the square root sign you have made a mistake.
- This mark would be gained for reaching $\sqrt{21}$.

Solutions

- Make sure you write down the non-rounded values from your calculator first.

Rounding

- This mark may be gained from correctly rounding a wrong answer.

NOTES: 4·7 page 43

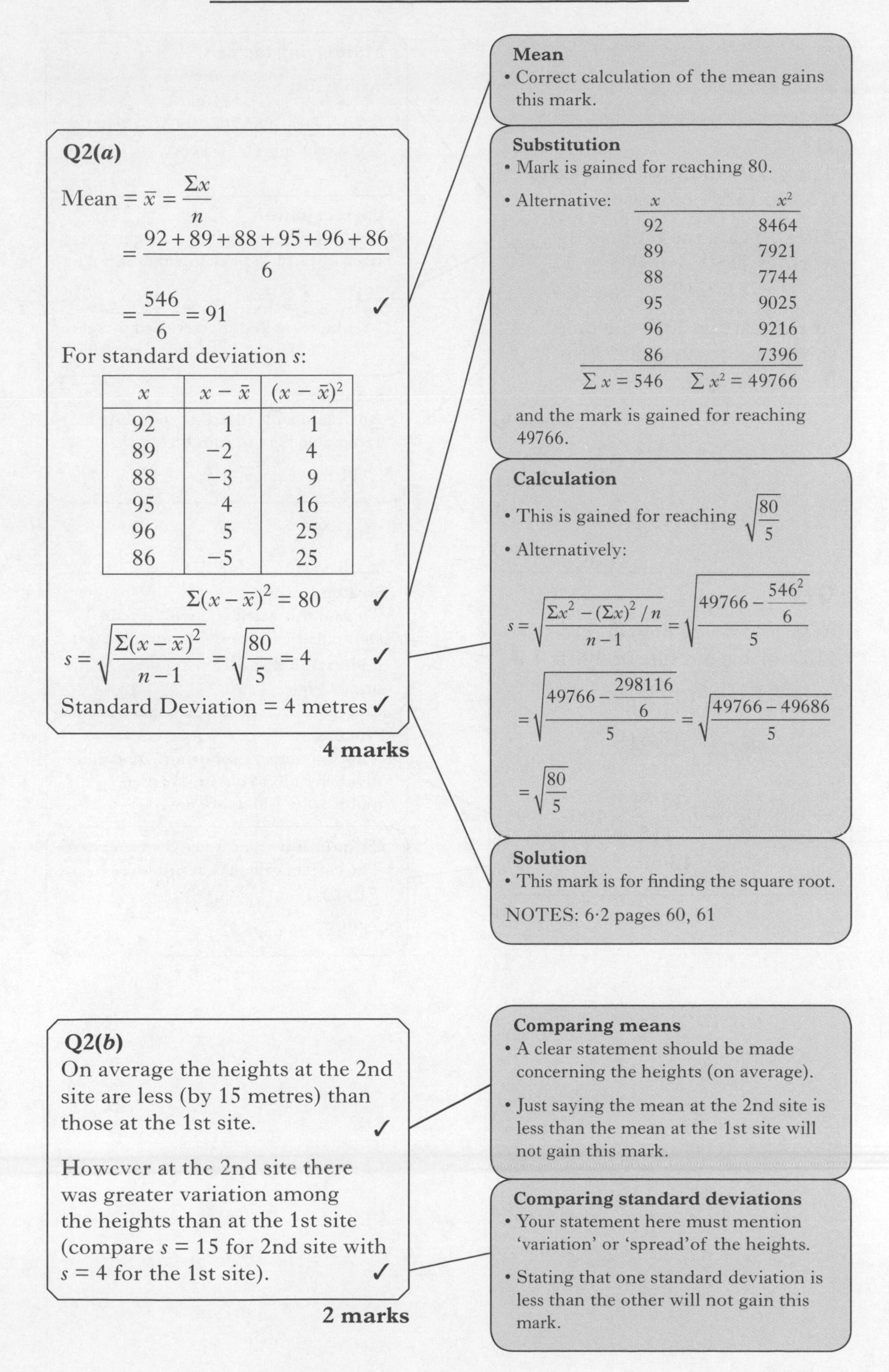

Q2(*a*)

$$\text{Mean} = \bar{x} = \frac{\Sigma x}{n}$$

$$= \frac{92+89+88+95+96+86}{6}$$

$$= \frac{546}{6} = 91 \quad ✓$$

For standard deviation *s*:

x	$x - \bar{x}$	$(x - \bar{x})^2$
92	1	1
89	−2	4
88	−3	9
95	4	16
96	5	25
86	−5	25

$$\Sigma(x-\bar{x})^2 = 80 \quad ✓$$

$$s = \sqrt{\frac{\Sigma(x-\bar{x})^2}{n-1}} = \sqrt{\frac{80}{5}} = 4 \quad ✓$$

Standard Deviation = 4 metres ✓

4 marks

Mean
- Correct calculation of the mean gains this mark.

Substitution
- Mark is gained for reaching 80.
- Alternative:

x	x^2
92	8464
89	7921
88	7744
95	9025
96	9216
86	7396
$\sum x = 546$	$\sum x^2 = 49766$

and the mark is gained for reaching 49766.

Calculation
- This is gained for reaching $\sqrt{\frac{80}{5}}$
- Alternatively:

$$s = \sqrt{\frac{\Sigma x^2 - (\Sigma x)^2/n}{n-1}} = \sqrt{\frac{49766 - \frac{546^2}{6}}{5}}$$

$$= \sqrt{\frac{49766 - \frac{298116}{6}}{5}} = \sqrt{\frac{49766 - 49686}{5}}$$

$$= \sqrt{\frac{80}{5}}$$

Solution
- This mark is for finding the square root.

NOTES: 6·2 pages 60, 61

Q2(*b*)

On average the heights at the 2nd site are less (by 15 metres) than those at the 1st site. ✓

However at the 2nd site there was greater variation among the heights than at the 1st site (compare $s = 15$ for 2nd site with $s = 4$ for the 1st site). ✓

2 marks

Comparing means
- A clear statement should be made concerning the heights (on average).
- Just saying the mean at the 2nd site is less than the mean at the 1st site will not gain this mark.

Comparing standard deviations
- Your statement here must mention 'variation' or 'spread' of the heights.
- Stating that one standard deviation is less than the other will not gain this mark.

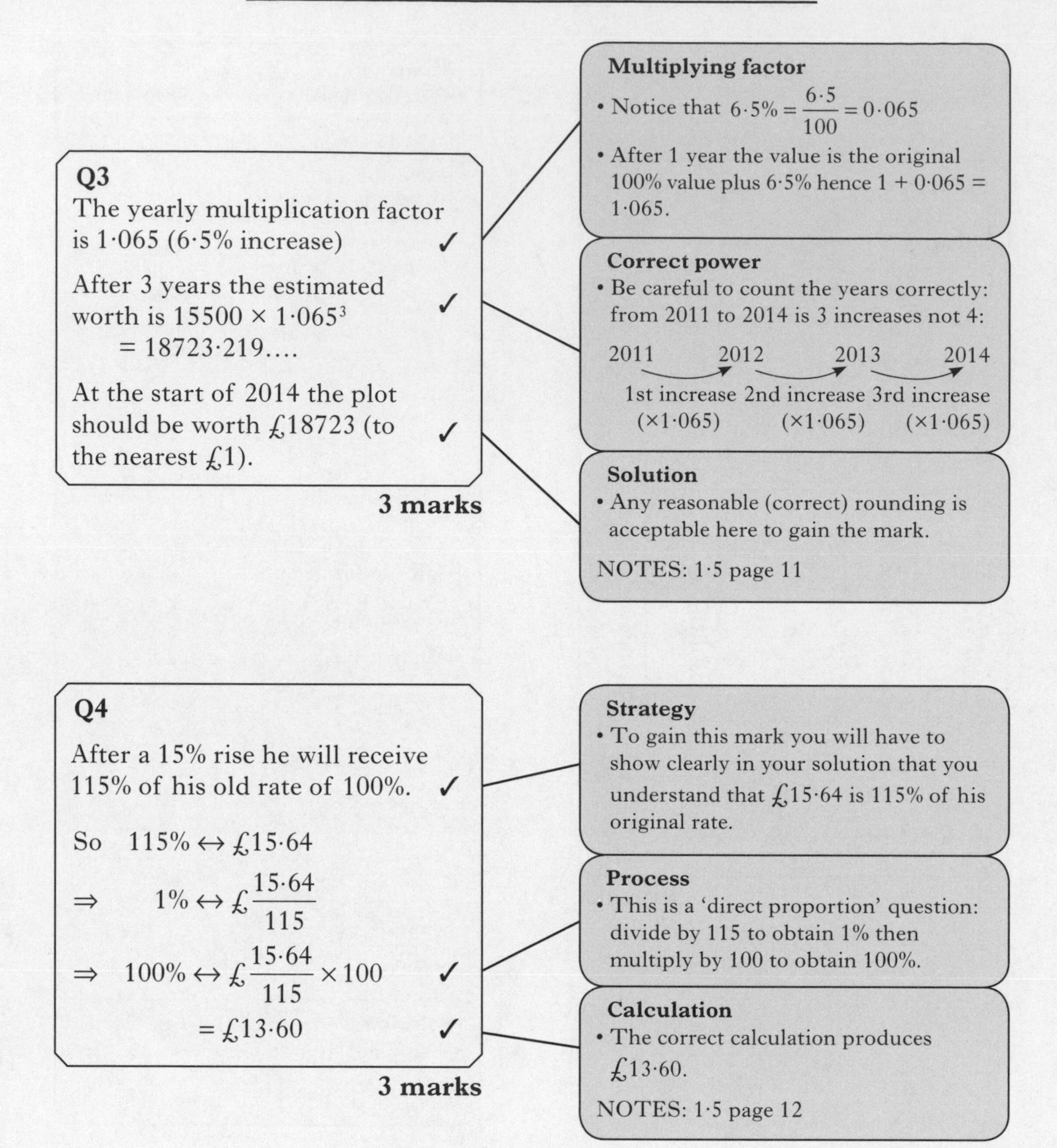
Q3
The yearly multiplication factor is 1·065 (6·5% increase) ✓
After 3 years the estimated worth is $15500 \times 1{\cdot}065^3 = 18723{\cdot}219....$ ✓
At the start of 2014 the plot should be worth £18723 (to the nearest £1). ✓
3 marks
Multiplying factor
• Notice that $6{\cdot}5\% = \frac{6{\cdot}5}{100} = 0{\cdot}065$
• After 1 year the value is the original 100% value plus 6·5% hence 1 + 0·065 = 1·065.
Correct power
• Be careful to count the years correctly: from 2011 to 2014 is 3 increases not 4:
2011 2012 2013 2014
1st increase (×1·065) 2nd increase (×1·065) 3rd increase (×1·065)
Solution
• Any reasonable (correct) rounding is acceptable here to gain the mark.
NOTES: 1·5 page 11
Q4
After a 15% rise he will receive 115% of his old rate of 100%. ✓
So 115% ↔ £15·64
$\Rightarrow \quad 1\% \leftrightarrow £\frac{15{\cdot}64}{115}$
$\Rightarrow \quad 100\% \leftrightarrow £\frac{15{\cdot}64}{115} \times 100$ ✓
= £13·60 ✓
3 marks
Strategy
• To gain this mark you will have to show clearly in your solution that you understand that £15·64 is 115% of his original rate.
Process
• This is a 'direct proportion' question: divide by 115 to obtain 1% then multiply by 100 to obtain 100%.
Calculation
• The correct calculation produces £13·60.
NOTES: 1·5 page 12

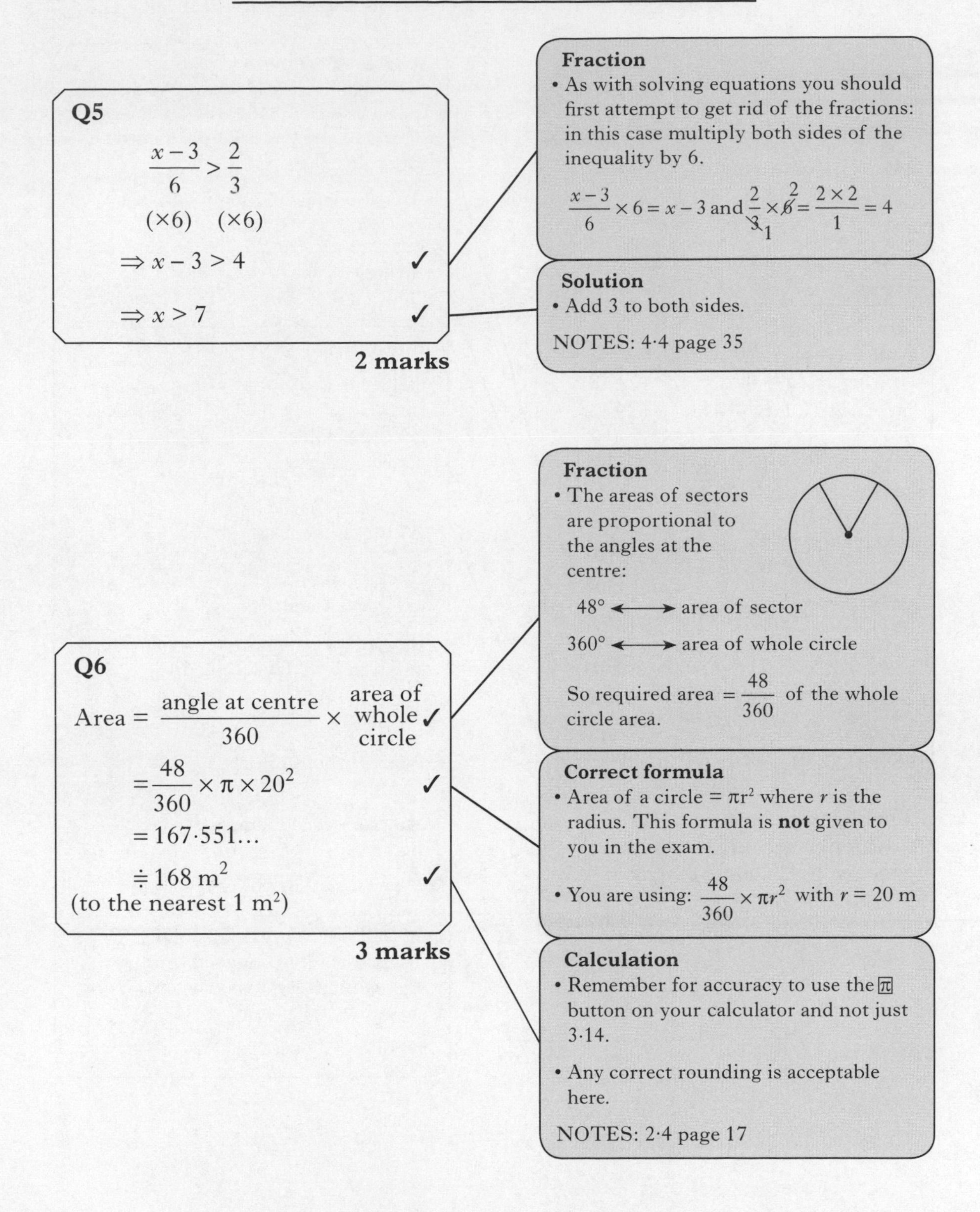

Q5

$$\frac{x-3}{6} > \frac{2}{3}$$
$(\times 6) \quad (\times 6)$

$\Rightarrow x - 3 > 4$ ✓

$\Rightarrow x > 7$ ✓

2 marks

Fraction
- As with solving equations you should first attempt to get rid of the fractions: in this case multiply both sides of the inequality by 6.

$$\frac{x-3}{6} \times 6 = x - 3 \text{ and } \frac{2}{3} \times 6 = \frac{2 \times 2}{1} = 4$$

Solution
- Add 3 to both sides.

NOTES: 4·4 page 35

Q6

$$\text{Area} = \frac{\text{angle at centre}}{360} \times \text{area of whole circle}$$ ✓

$$= \frac{48}{360} \times \pi \times 20^2$$ ✓

$$= 167{\cdot}551\ldots$$

$$\doteqdot 168 \text{ m}^2$$ ✓
(to the nearest 1 m^2)

3 marks

Fraction
- The areas of sectors are proportional to the angles at the centre:

48° ⟷ area of sector

360° ⟷ area of whole circle

So required area $= \frac{48}{360}$ of the whole circle area.

Correct formula
- Area of a circle $= \pi r^2$ where r is the radius. This formula is **not** given to you in the exam.
- You are using: $\frac{48}{360} \times \pi r^2$ with $r = 20$ m

Calculation
- Remember for accuracy to use the π button on your calculator and not just 3·14.
- Any correct rounding is acceptable here.

NOTES: 2·4 page 17

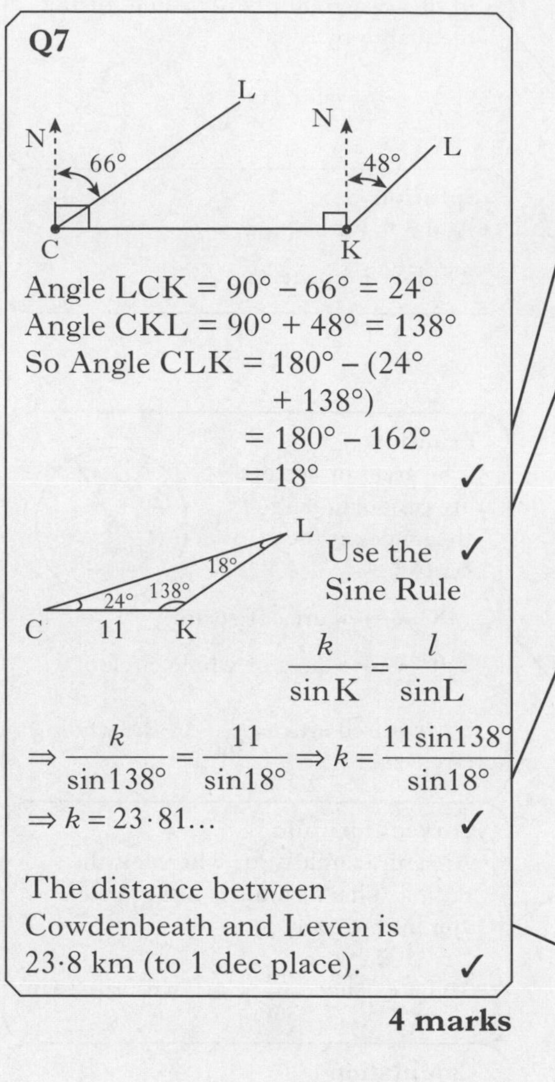

Q7

Angle LCK = 90° – 66° = 24°
Angle CKL = 90° + 48° = 138°
So Angle CLK = 180° – (24° + 138°)
= 180° – 162°
= 18° ✓

Use the Sine Rule ✓

$$\frac{k}{\sin K} = \frac{l}{\sin L}$$

$$\Rightarrow \frac{k}{\sin 138°} = \frac{11}{\sin 18°} \Rightarrow k = \frac{11 \sin 138°}{\sin 18°}$$

$\Rightarrow k = 23{\cdot}81\ldots$ ✓

The distance between Cowdenbeath and Leven is 23·8 km (to 1 dec place). ✓

4 marks

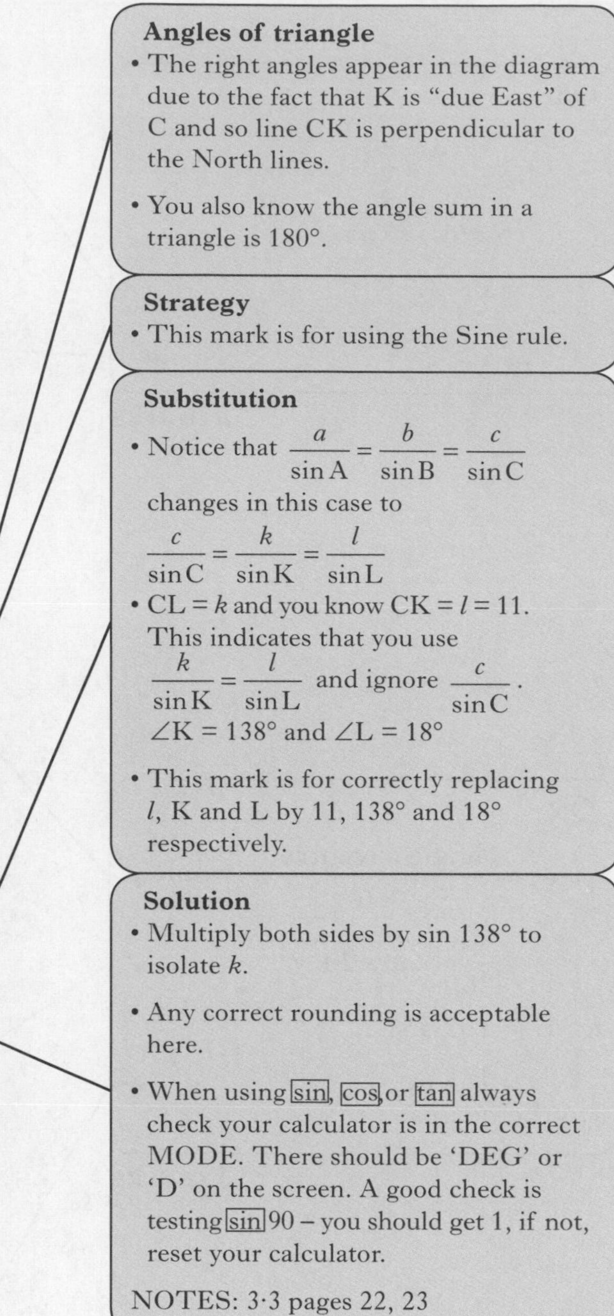

Angles of triangle

- The right angles appear in the diagram due to the fact that K is "due East" of C and so line CK is perpendicular to the North lines.
- You also know the angle sum in a triangle is 180°.

Strategy

- This mark is for using the Sine rule.

Substitution

- Notice that $\frac{a}{\sin A} = \frac{b}{\sin B} = \frac{c}{\sin C}$ changes in this case to $\frac{c}{\sin C} = \frac{k}{\sin K} = \frac{l}{\sin L}$
- CL = k and you know CK = l = 11. This indicates that you use $\frac{k}{\sin K} = \frac{l}{\sin L}$ and ignore $\frac{c}{\sin C}$.
  ∠K = 138° and ∠L = 18°
- This mark is for correctly replacing l, K and L by 11, 138° and 18° respectively.

Solution

- Multiply both sides by sin 138° to isolate k.
- Any correct rounding is acceptable here.
- When using [sin], [cos], or [tan] always check your calculator is in the correct MODE. There should be 'DEG' or 'D' on the screen. A good check is testing [sin] 90 – you should get 1, if not, reset your calculator.

NOTES: 3·3 pages 22, 23

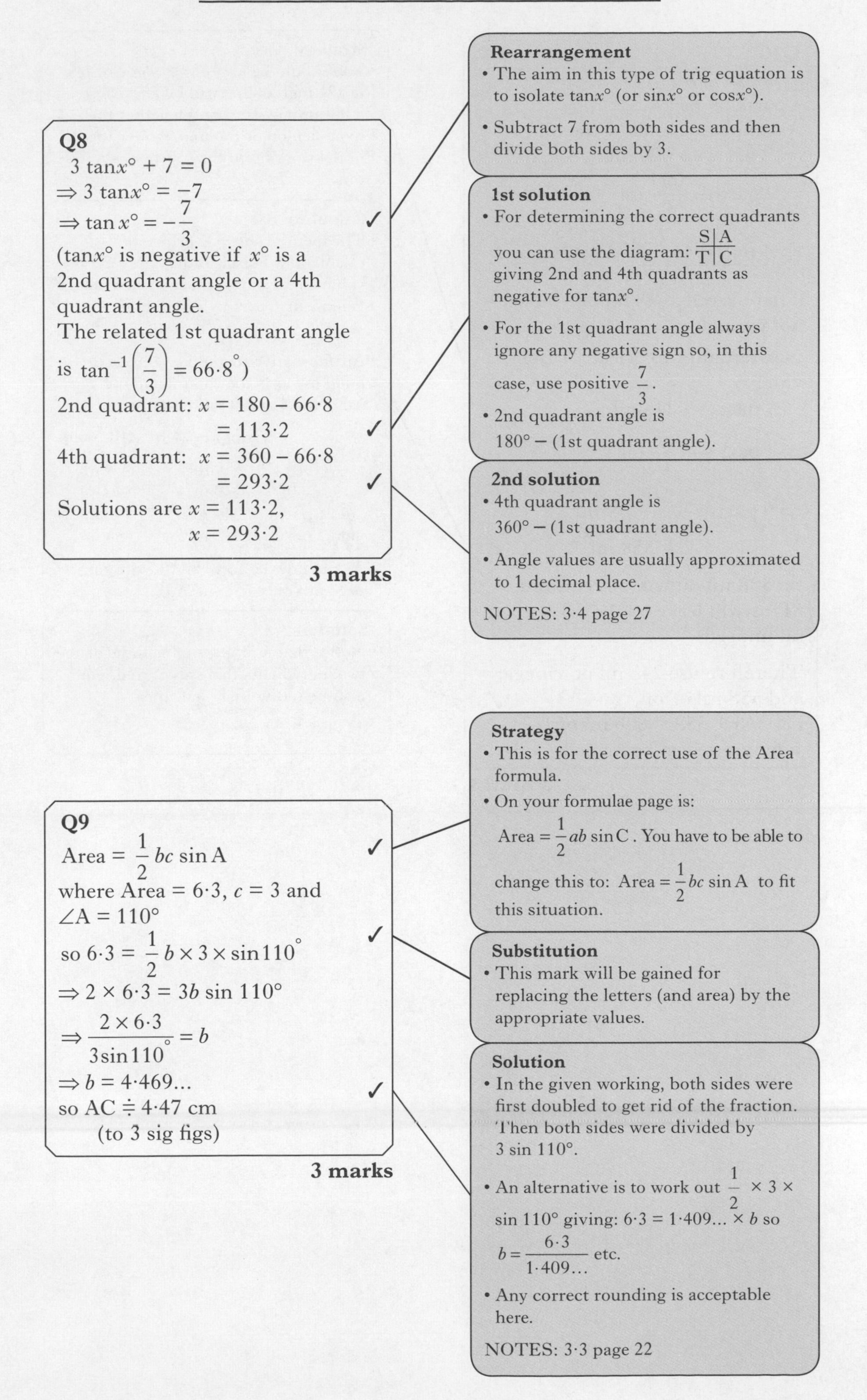

Q8

$3\tan x^\circ + 7 = 0$

$\Rightarrow 3\tan x^\circ = -7$

$\Rightarrow \tan x^\circ = -\frac{7}{3}$ ✓

($\tan x^\circ$ is negative if x° is a 2nd quadrant angle or a 4th quadrant angle.
The related 1st quadrant angle is $\tan^{-1}\left(\frac{7}{3}\right) = 66{\cdot}8^\circ$)

2nd quadrant: $x = 180 - 66{\cdot}8$
$= 113{\cdot}2$ ✓

4th quadrant: $x = 360 - 66{\cdot}8$
$= 293{\cdot}2$ ✓

Solutions are $x = 113{\cdot}2$,
$x = 293{\cdot}2$

3 marks

Rearrangement
- The aim in this type of trig equation is to isolate $\tan x^\circ$ (or $\sin x^\circ$ or $\cos x^\circ$).
- Subtract 7 from both sides and then divide both sides by 3.

1st solution
- For determining the correct quadrants you can use the diagram: $\begin{array}{c|c} S & A \\ \hline T & C \end{array}$ giving 2nd and 4th quadrants as negative for $\tan x^\circ$.
- For the 1st quadrant angle always ignore any negative sign so, in this case, use positive $\frac{7}{3}$.
- 2nd quadrant angle is $180^\circ - $ (1st quadrant angle).

2nd solution
- 4th quadrant angle is $360^\circ - $ (1st quadrant angle).
- Angle values are usually approximated to 1 decimal place.

NOTES: 3·4 page 27

Q9

$\text{Area} = \frac{1}{2}bc\sin A$ ✓

where Area $= 6{\cdot}3$, $c = 3$ and $\angle A = 110^\circ$

so $6{\cdot}3 = \frac{1}{2} b \times 3 \times \sin 110^\circ$ ✓

$\Rightarrow 2 \times 6{\cdot}3 = 3b\sin 110^\circ$

$\Rightarrow \frac{2 \times 6{\cdot}3}{3\sin 110^\circ} = b$

$\Rightarrow b = 4{\cdot}469\ldots$ ✓

so AC $\doteqdot 4{\cdot}47$ cm
(to 3 sig figs)

3 marks

Strategy
- This is for the correct use of the Area formula.
- On your formulae page is: $\text{Area} = \frac{1}{2}ab\sin C$. You have to be able to change this to: $\text{Area} = \frac{1}{2}bc\sin A$ to fit this situation.

Substitution
- This mark will be gained for replacing the letters (and area) by the appropriate values.

Solution
- In the given working, both sides were first doubled to get rid of the fraction. Then both sides were divided by $3\sin 110^\circ$.
- An alternative is to work out $\frac{1}{2} \times 3 \times \sin 110^\circ$ giving: $6{\cdot}3 = 1{\cdot}409\ldots \times b$ so $b = \frac{6{\cdot}3}{1{\cdot}409\ldots}$ etc.
- Any correct rounding is acceptable here.

NOTES: 3·3 page 22

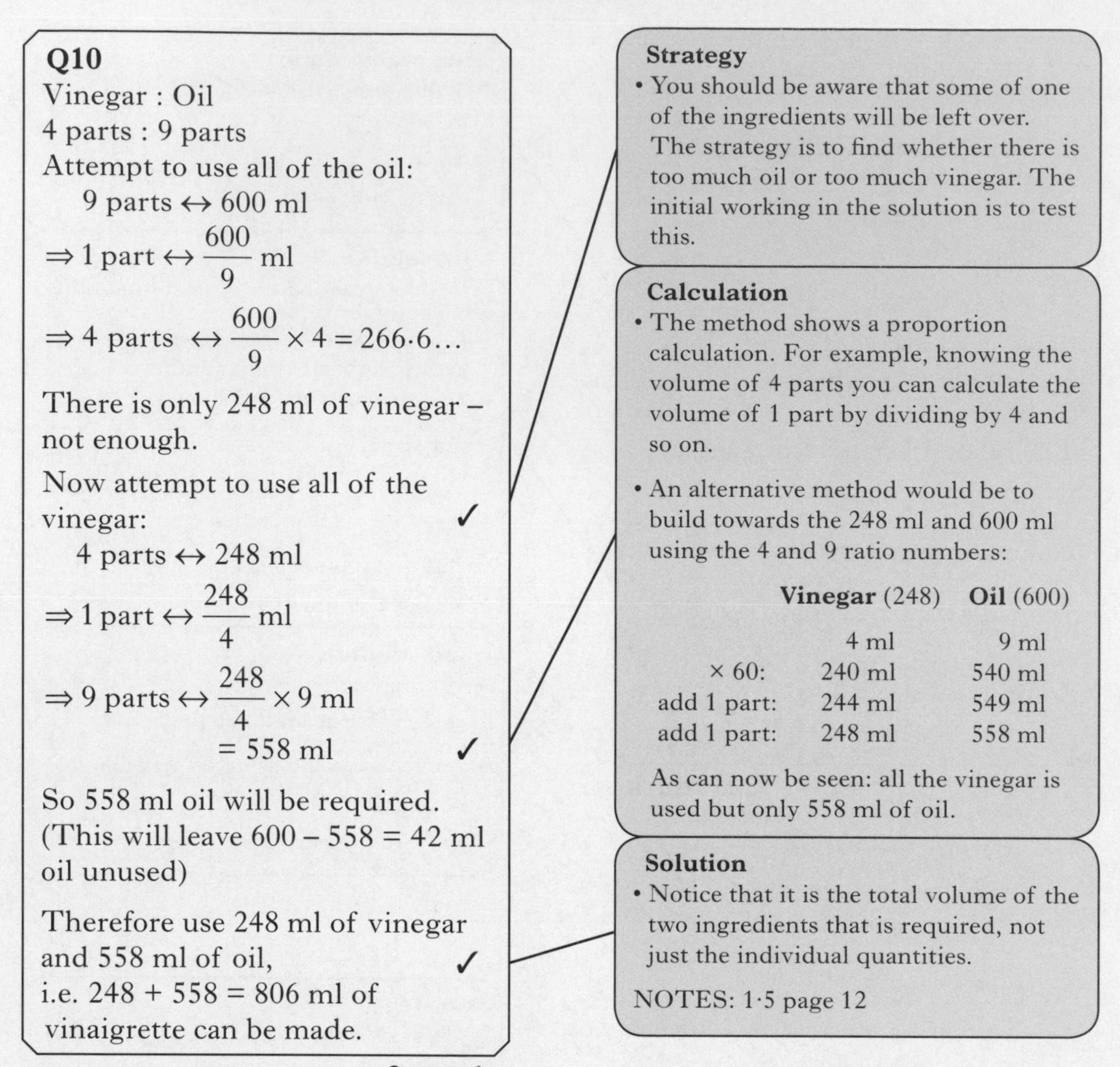

Q10

Vinegar : Oil

4 parts : 9 parts

Attempt to use all of the oil:

$9 \text{ parts} \leftrightarrow 600 \text{ ml}$

$\Rightarrow 1 \text{ part} \leftrightarrow \frac{600}{9} \text{ ml}$

$\Rightarrow 4 \text{ parts} \leftrightarrow \frac{600}{9} \times 4 = 266{\cdot}6\ldots$

There is only 248 ml of vinegar – not enough.

Now attempt to use all of the vinegar: ✓

$4 \text{ parts} \leftrightarrow 248 \text{ ml}$

$\Rightarrow 1 \text{ part} \leftrightarrow \frac{248}{4} \text{ ml}$

$\Rightarrow 9 \text{ parts} \leftrightarrow \frac{248}{4} \times 9 \text{ ml}$

$= 558 \text{ ml}$ ✓

So 558 ml oil will be required. (This will leave 600 – 558 = 42 ml oil unused)

Therefore use 248 ml of vinegar and 558 ml of oil, ✓
i.e. 248 + 558 = 806 ml of vinaigrette can be made.

3 marks

Strategy

- You should be aware that some of one of the ingredients will be left over. The strategy is to find whether there is too much oil or too much vinegar. The initial working in the solution is to test this.

Calculation

- The method shows a proportion calculation. For example, knowing the volume of 4 parts you can calculate the volume of 1 part by dividing by 4 and so on.
- An alternative method would be to build towards the 248 ml and 600 ml using the 4 and 9 ratio numbers:

	Vinegar (248)	**Oil** (600)
	4 ml	9 ml
× 60:	240 ml	540 ml
add 1 part:	244 ml	549 ml
add 1 part:	248 ml	558 ml

As can now be seen: all the vinegar is used but only 558 ml of oil.

Solution

- Notice that it is the total volume of the two ingredients that is required, not just the individual quantities.

NOTES: 1·5 page 12

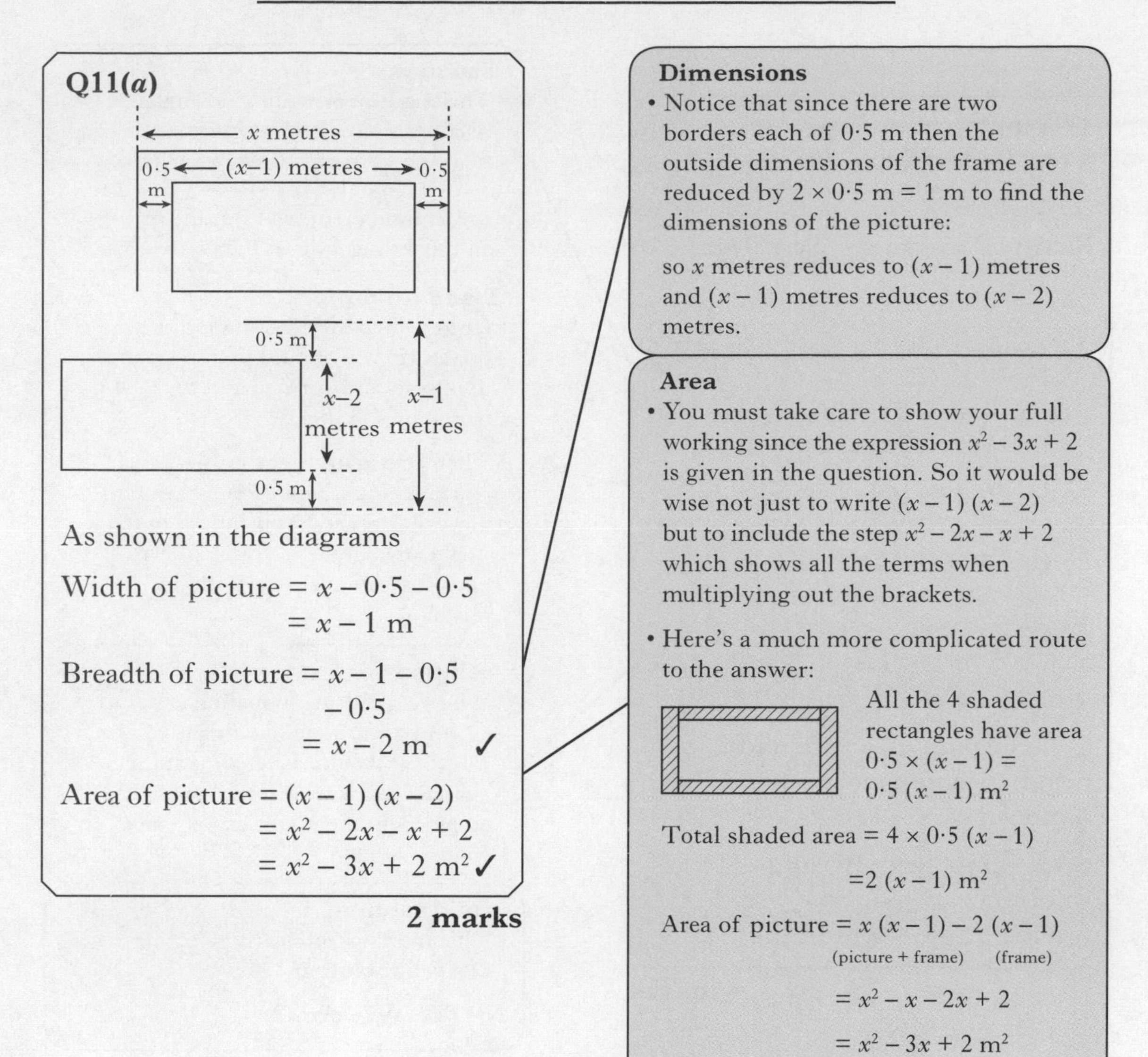

Q11(*a*)

As shown in the diagrams

Width of picture $= x - 0{\cdot}5 - 0{\cdot}5$
$= x - 1$ m

Breadth of picture $= x - 1 - 0{\cdot}5 - 0{\cdot}5$
$= x - 2$ m ✓

Area of picture $= (x - 1)(x - 2)$
$= x^2 - 2x - x + 2$
$= x^2 - 3x + 2$ m^2 ✓

2 marks

Dimensions

- Notice that since there are two borders each of 0·5 m then the outside dimensions of the frame are reduced by $2 \times 0{\cdot}5$ m $= 1$ m to find the dimensions of the picture:

  so x metres reduces to $(x - 1)$ metres and $(x - 1)$ metres reduces to $(x - 2)$ metres.

Area

- You must take care to show your full working since the expression $x^2 - 3x + 2$ is given in the question. So it would be wise not just to write $(x - 1)(x - 2)$ but to include the step $x^2 - 2x - x + 2$ which shows all the terms when multiplying out the brackets.
- Here's a much more complicated route to the answer:

  All the 4 shaded rectangles have area $0{\cdot}5 \times (x - 1) = 0{\cdot}5\,(x - 1)$ m^2

  Total shaded area $= 4 \times 0{\cdot}5\,(x - 1)$
  $= 2\,(x - 1)$ m^2

  Area of picture $= x\,(x - 1) - 2\,(x - 1)$
  (picture + frame) (frame)
  $= x^2 - x - 2x + 2$
  $= x^2 - 3x + 2$ m^2

NOTES: 4·2 pages 30, 31

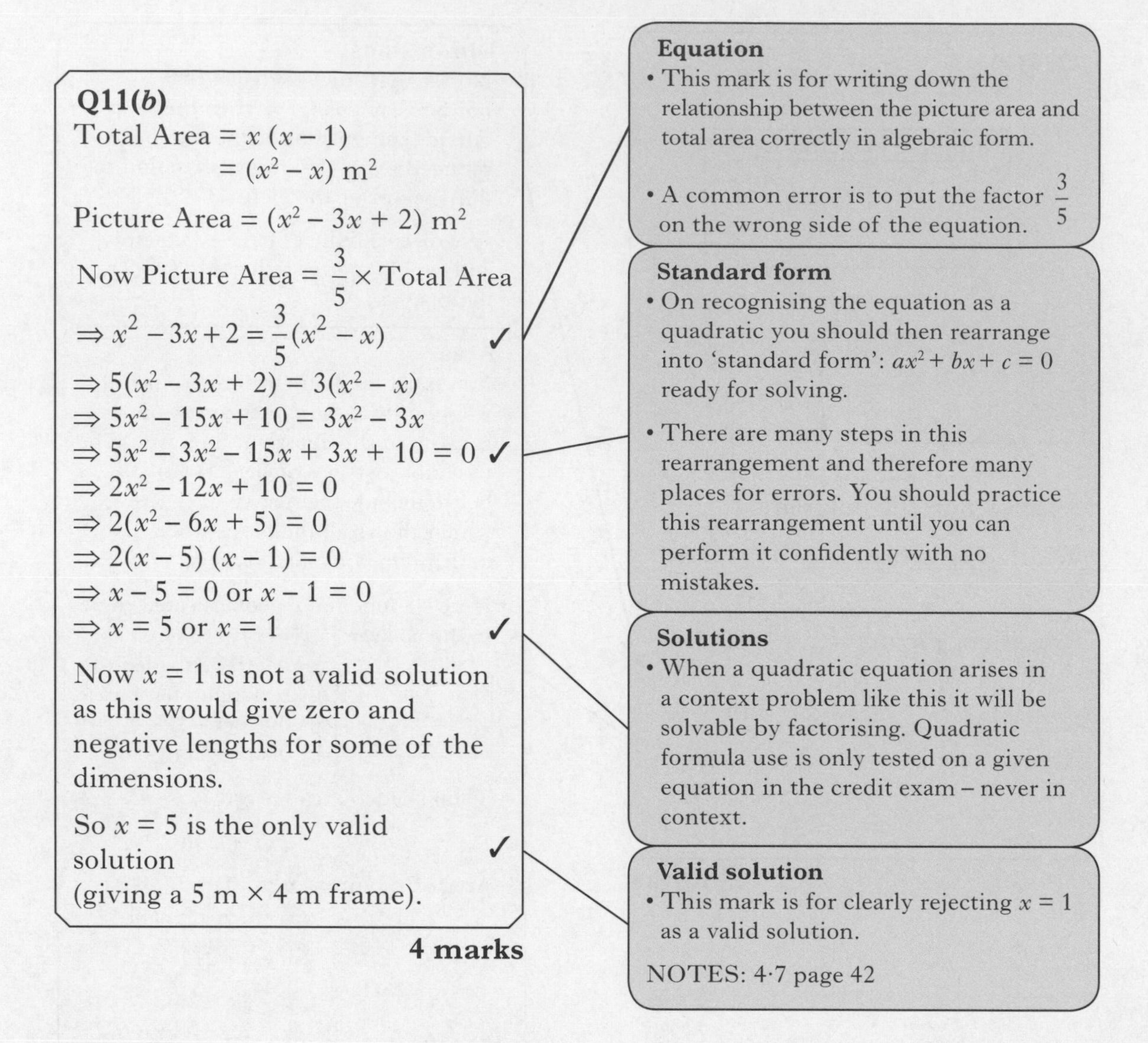

Q11(b)
Total Area = $x(x-1)$
$= (x^2 - x)\ \text{m}^2$
Picture Area = $(x^2 - 3x + 2)\ \text{m}^2$
Now Picture Area = $\frac{3}{5} \times$ Total Area
$\Rightarrow x^2 - 3x + 2 = \frac{3}{5}(x^2 - x)$ ✓
$\Rightarrow 5(x^2 - 3x + 2) = 3(x^2 - x)$
$\Rightarrow 5x^2 - 15x + 10 = 3x^2 - 3x$
$\Rightarrow 5x^2 - 3x^2 - 15x + 3x + 10 = 0$ ✓
$\Rightarrow 2x^2 - 12x + 10 = 0$
$\Rightarrow 2(x^2 - 6x + 5) = 0$
$\Rightarrow 2(x - 5)(x - 1) = 0$
$\Rightarrow x - 5 = 0$ or $x - 1 = 0$
$\Rightarrow x = 5$ or $x = 1$ ✓
Now $x = 1$ is not a valid solution as this would give zero and negative lengths for some of the dimensions.
So $x = 5$ is the only valid solution (giving a 5 m × 4 m frame). ✓
4 marks
Equation
• This mark is for writing down the relationship between the picture area and total area correctly in algebraic form.
• A common error is to put the factor $\frac{3}{5}$ on the wrong side of the equation.
Standard form
• On recognising the equation as a quadratic you should then rearrange into 'standard form': $ax^2 + bx + c = 0$ ready for solving.
• There are many steps in this rearrangement and therefore many places for errors. You should practice this rearrangement until you can perform it confidently with no mistakes.
Solutions
• When a quadratic equation arises in a context problem like this it will be solvable by factorising. Quadratic formula use is only tested on a given equation in the credit exam – never in context.
Valid solution
• This mark is for clearly rejecting $x = 1$ as a valid solution.
NOTES: 4·7 page 42

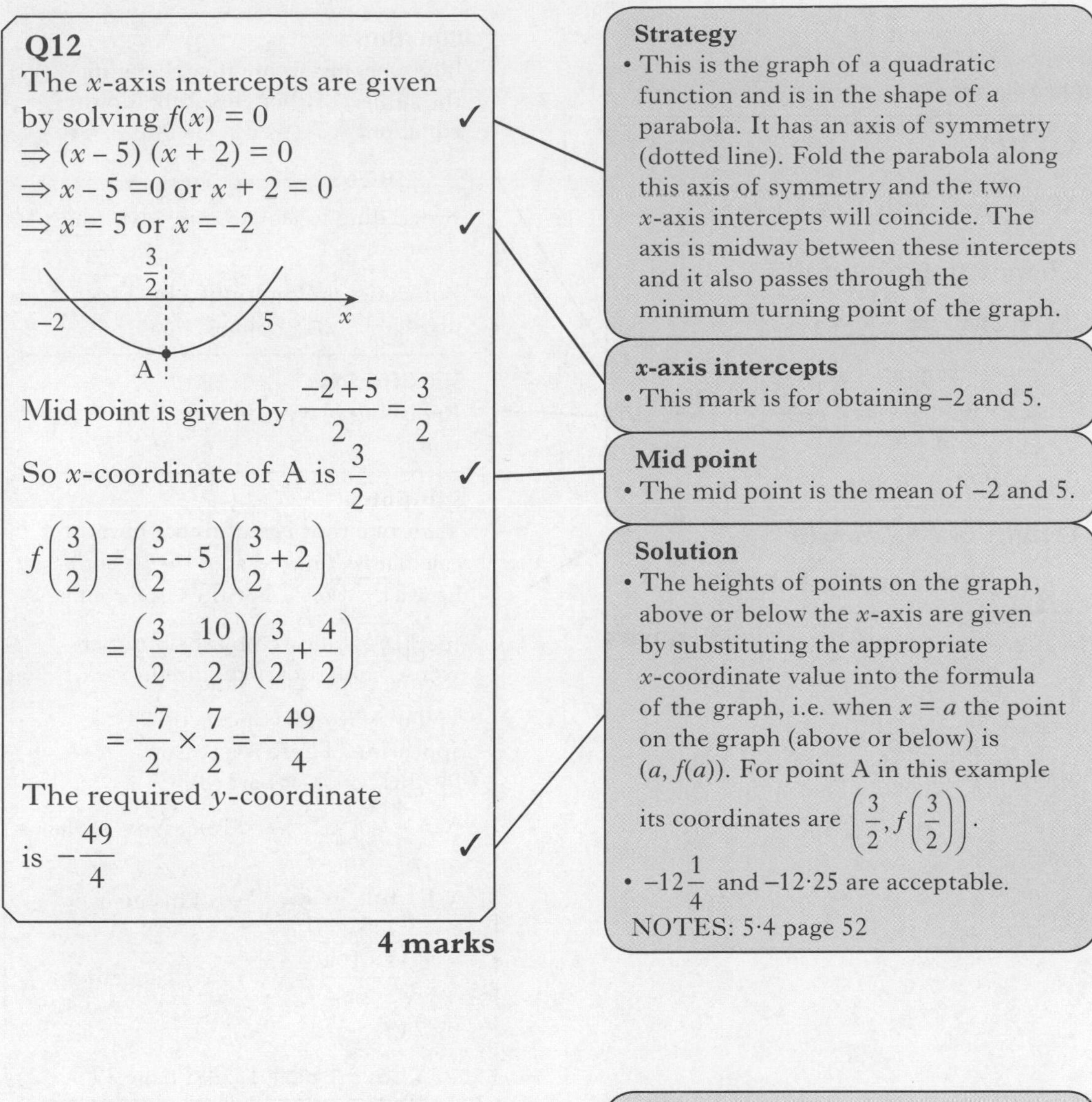

Q12

The x-axis intercepts are given by solving $f(x) = 0$ ✓

$\Rightarrow (x - 5)(x + 2) = 0$

$\Rightarrow x - 5 = 0$ or $x + 2 = 0$

$\Rightarrow x = 5$ or $x = -2$ ✓

Mid point is given by $\frac{-2+5}{2} = \frac{3}{2}$

So x-coordinate of A is $\frac{3}{2}$ ✓

$$f\left(\frac{3}{2}\right) = \left(\frac{3}{2} - 5\right)\left(\frac{3}{2} + 2\right)$$

$$= \left(\frac{3}{2} - \frac{10}{2}\right)\left(\frac{3}{2} + \frac{4}{2}\right)$$

$$= \frac{-7}{2} \times \frac{7}{2} = -\frac{49}{4}$$

The required y-coordinate is $-\frac{49}{4}$ ✓

4 marks

Strategy

- This is the graph of a quadratic function and is in the shape of a parabola. It has an axis of symmetry (dotted line). Fold the parabola along this axis of symmetry and the two x-axis intercepts will coincide. The axis is midway between these intercepts and it also passes through the minimum turning point of the graph.

***x*-axis intercepts**

- This mark is for obtaining −2 and 5.

Mid point

- The mid point is the mean of −2 and 5.

Solution

- The heights of points on the graph, above or below the x-axis are given by substituting the appropriate x-coordinate value into the formula of the graph, i.e. when $x = a$ the point on the graph (above or below) is $(a, f(a))$. For point A in this example its coordinates are $\left(\frac{3}{2}, f\left(\frac{3}{2}\right)\right)$.
- $-12\frac{1}{4}$ and −12·25 are acceptable.

NOTES: 5·4 page 52

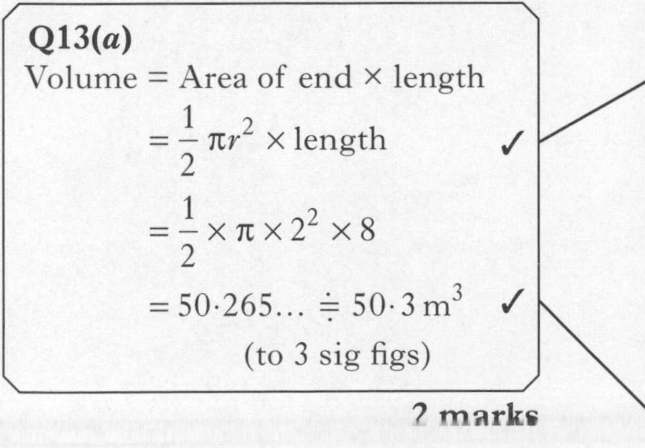

Q13(*a*)

Volume = Area of end × length

$= \frac{1}{2}\pi r^2 \times \text{length}$ ✓

$= \frac{1}{2} \times \pi \times 2^2 \times 8$

$= 50{\cdot}265\ldots \doteqdot 50{\cdot}3\ \text{m}^3$ ✓

(to 3 sig figs)

2 marks

Formula

- Volume of a Prism = Area of end × length. These ‘row covers’ are prisms with ends that are semicircles.
- You may know the volume of a cylinder formula: Volume = $\pi r^2 h$. In this case h is the length and you need to halve the result.

Solution

- Remember to use the π button on your calculator not 3·14.
- Any correct rounding is acceptable here.

NOTES: 2·1 page 14

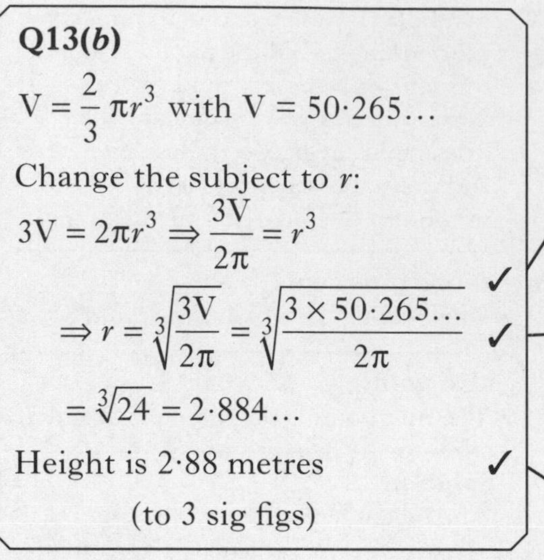

Q13(*b*)

$V = \frac{2}{3}\pi r^3$ with $V = 50{\cdot}265\ldots$

Change the subject to r:

$3V = 2\pi r^3 \Rightarrow \frac{3V}{2\pi} = r^3$ ✓

$\Rightarrow r = \sqrt[3]{\frac{3V}{2\pi}} = \sqrt[3]{\frac{3 \times 50{\cdot}265\ldots}{2\pi}}$ ✓

$= \sqrt[3]{24} = 2{\cdot}884\ldots$

Height is 2·88 metres (to 3 sig figs) ✓

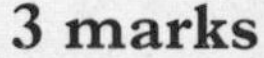

3 marks

Equation

- The working shows the 'changing the subject' although setting up the equation:

  $50{\cdot}265\cdots = \frac{2}{3}\pi r^3$ and

  proceeding to isolate r^3 involves the same steps.
- Both sides are multiplied by 3 then divided by 2π to isolate r^3.

Substitution

- Replacing V with 50·265... gains the mark.

Solution

- The cube root is required. On most calculators there is a [$\sqrt[x]{\ }$] key. (Some have a [$\sqrt[3]{\ }$] key). For the cube root key in: [3] [$\sqrt[x]{\ }$] followed by the number whose cube root is required.
- You may wonder at exactly 24 appearing. There is a reason for this. Look at part (*a*):

  $V = \frac{1}{2} \times \pi \times 2^2 \times 8 = 16\pi$. Now replace V by 16π in $r = \sqrt[3]{\frac{3V}{2\pi}}$. This gives

  $r = \sqrt[3]{\frac{3 \times 16\pi}{2\pi}} = \sqrt[3]{24}$ after cancelling by 2π.

NOTES: 2·1 page 13, 4·5 page 39

Practice Exam F: Paper 1 Worked Answers

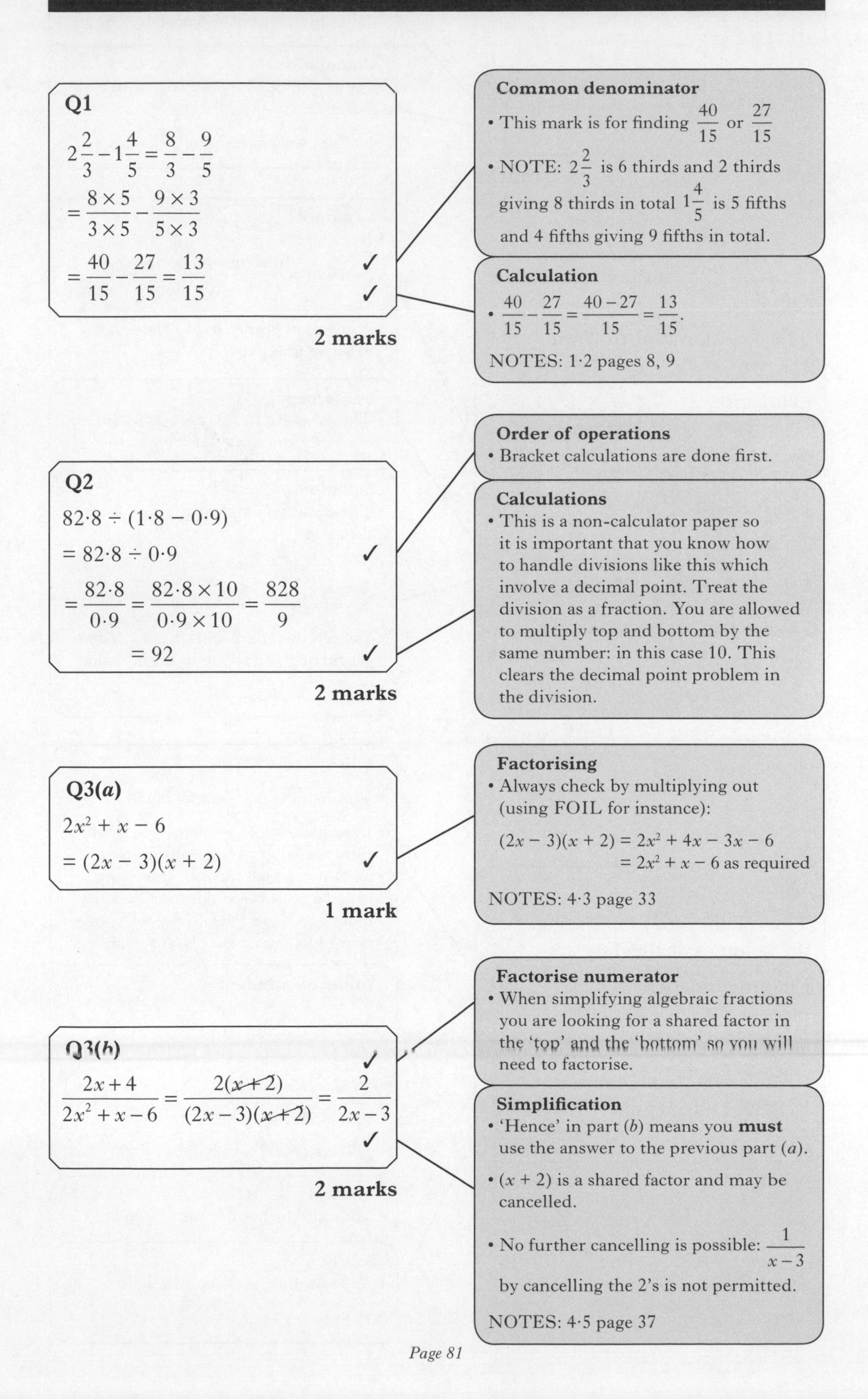

Q1

$$2\frac{2}{3} - 1\frac{4}{5} = \frac{8}{3} - \frac{9}{5}$$

$$= \frac{8 \times 5}{3 \times 5} - \frac{9 \times 3}{5 \times 3}$$

$$= \frac{40}{15} - \frac{27}{15} = \frac{13}{15}$$ ✓ ✓

2 marks

Common denominator

- This mark is for finding $\frac{40}{15}$ or $\frac{27}{15}$
- NOTE: $2\frac{2}{3}$ is 6 thirds and 2 thirds giving 8 thirds in total $1\frac{4}{5}$ is 5 fifths and 4 fifths giving 9 fifths in total.

Calculation

- $\frac{40}{15} - \frac{27}{15} = \frac{40-27}{15} = \frac{13}{15}$.

NOTES: 1·2 pages 8, 9

Q2

$82{\cdot}8 \div (1{\cdot}8 - 0{\cdot}9)$

$= 82{\cdot}8 \div 0{\cdot}9$ ✓

$$= \frac{82{\cdot}8}{0{\cdot}9} = \frac{82{\cdot}8 \times 10}{0{\cdot}9 \times 10} = \frac{828}{9}$$

$= 92$ ✓

2 marks

Order of operations

- Bracket calculations are done first.

Calculations

- This is a non-calculator paper so it is important that you know how to handle divisions like this which involve a decimal point. Treat the division as a fraction. You are allowed to multiply top and bottom by the same number: in this case 10. This clears the decimal point problem in the division.

Q3(*a*)

$2x^2 + x - 6$

$= (2x - 3)(x + 2)$ ✓

1 mark

Factorising

- Always check by multiplying out (using FOIL for instance):

$(2x - 3)(x + 2) = 2x^2 + 4x - 3x - 6$
$= 2x^2 + x - 6$ as required

NOTES: 4·3 page 33

Q3(*b*)

$$\frac{2x+4}{2x^2+x-6} = \frac{2(x+2)}{(2x-3)(x+2)} = \frac{2}{2x-3}$$ ✓ ✓

2 marks

Factorise numerator

- When simplifying algebraic fractions you are looking for a shared factor in the 'top' and the 'bottom' so you will need to factorise.

Simplification

- 'Hence' in part (*b*) means you **must** use the answer to the previous part (*a*).
- $(x + 2)$ is a shared factor and may be cancelled.
- No further cancelling is possible: $\frac{1}{x-3}$ by cancelling the 2's is not permitted.

NOTES: 4·5 page 37

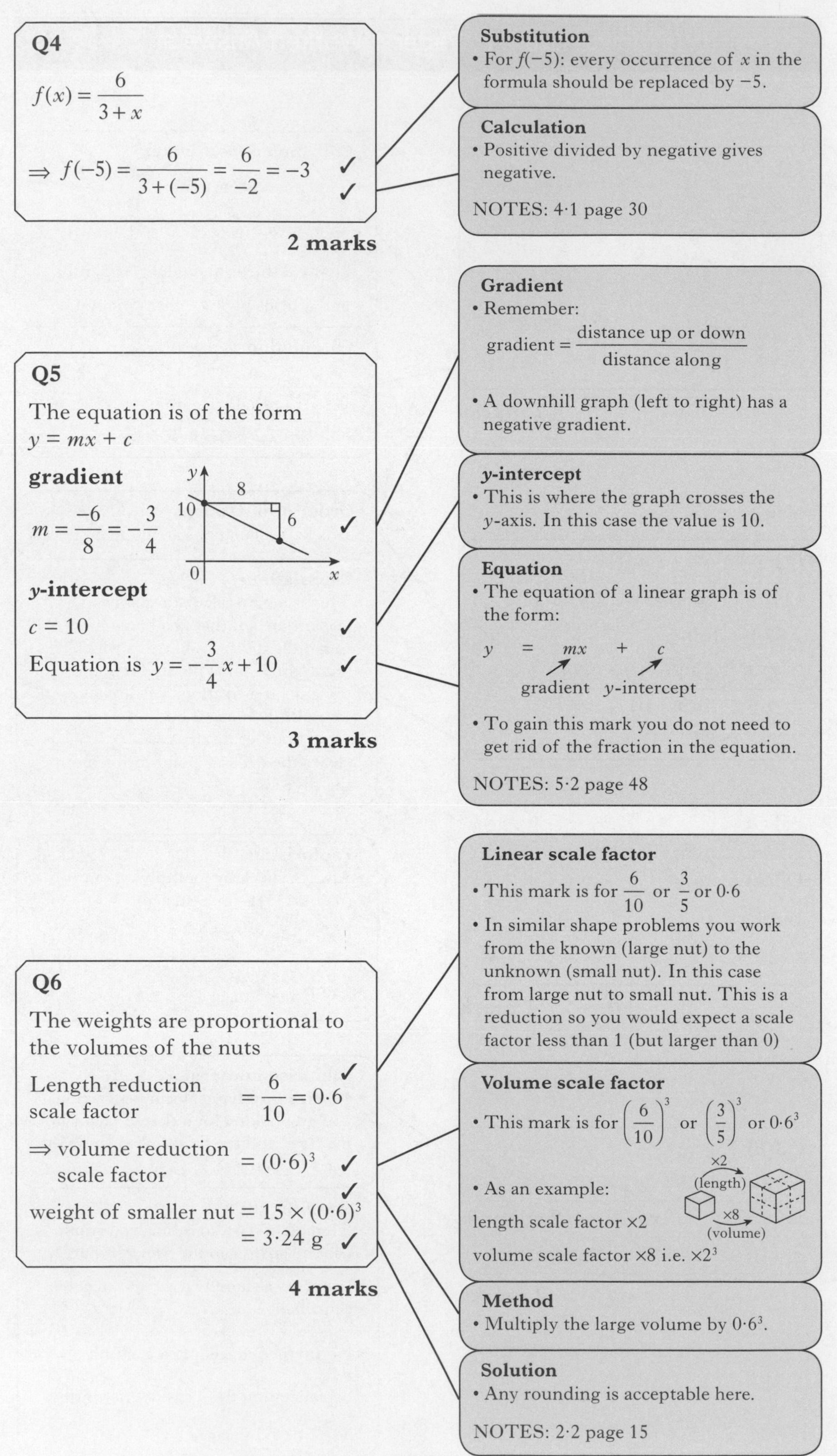

Q4

$f(x) = \dfrac{6}{3+x}$

$\Rightarrow f(-5) = \dfrac{6}{3+(-5)} = \dfrac{6}{-2} = -3$ ✓ ✓

2 marks

Substitution
- For $f(-5)$: every occurrence of x in the formula should be replaced by -5.

Calculation
- Positive divided by negative gives negative.

NOTES: 4·1 page 30

Q5

The equation is of the form $y = mx + c$

gradient

$m = \dfrac{-6}{8} = -\dfrac{3}{4}$ ✓

***y*-intercept**

$c = 10$ ✓

Equation is $y = -\dfrac{3}{4}x + 10$ ✓

3 marks

Gradient
- Remember:

$$\text{gradient} = \frac{\text{distance up or down}}{\text{distance along}}$$

- A downhill graph (left to right) has a negative gradient.

***y*-intercept**
- This is where the graph crosses the y-axis. In this case the value is 10.

Equation
- The equation of a linear graph is of the form:

$y = mx + c$ (m: gradient, c: y-intercept)

- To gain this mark you do not need to get rid of the fraction in the equation.

NOTES: 5·2 page 48

Q6

The weights are proportional to the volumes of the nuts

Length reduction scale factor $= \dfrac{6}{10} = 0{\cdot}6$ ✓

$\Rightarrow$ volume reduction scale factor $= (0{\cdot}6)^3$ ✓

weight of smaller nut $= 15 \times (0{\cdot}6)^3$ ✓
$= 3{\cdot}24$ g ✓

4 marks

Linear scale factor
- This mark is for $\dfrac{6}{10}$ or $\dfrac{3}{5}$ or $0{\cdot}6$
- In similar shape problems you work from the known (large nut) to the unknown (small nut). In this case from large nut to small nut. This is a reduction so you would expect a scale factor less than 1 (but larger than 0)

Volume scale factor
- This mark is for $\left(\dfrac{6}{10}\right)^3$ or $\left(\dfrac{3}{5}\right)^3$ or $0{\cdot}6^3$
- As an example:

length scale factor ×2
volume scale factor ×8 i.e. $\times 2^3$

Method
- Multiply the large volume by $0{\cdot}6^3$.

Solution
- Any rounding is acceptable here.

NOTES: 2·2 page 15

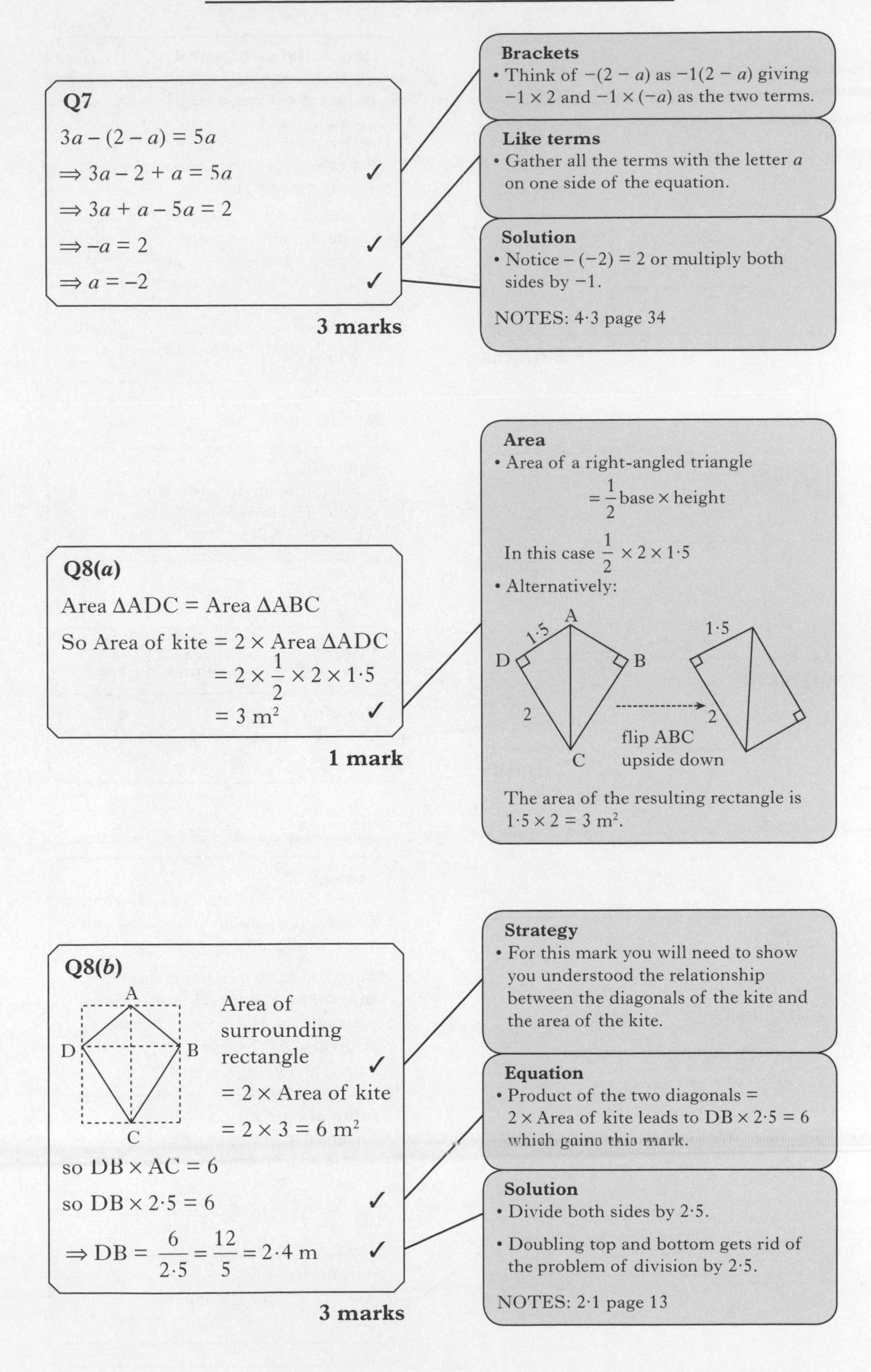

Q7

$3a - (2 - a) = 5a$

$\Rightarrow 3a - 2 + a = 5a$ ✓

$\Rightarrow 3a + a - 5a = 2$

$\Rightarrow -a = 2$ ✓

$\Rightarrow a = -2$ ✓

3 marks

Brackets

- Think of $-(2 - a)$ as $-1(2 - a)$ giving -1×2 and $-1 \times (-a)$ as the two terms.

Like terms

- Gather all the terms with the letter a on one side of the equation.

Solution

- Notice $-(-2) = 2$ or multiply both sides by -1.

NOTES: 4·3 page 34

Q8(*a*)

Area ΔADC = Area ΔABC

So Area of kite = 2 × Area ΔADC

$= 2 \times \frac{1}{2} \times 2 \times 1{\cdot}5$

$= 3 \text{ m}^2$ ✓

1 mark

Area

- Area of a right-angled triangle

$$= \frac{1}{2} \text{base} \times \text{height}$$

In this case $\frac{1}{2} \times 2 \times 1{\cdot}5$

- Alternatively:

The area of the resulting rectangle is $1{\cdot}5 \times 2 = 3 \text{ m}^2$.

Q8(*b*)

Area of surrounding rectangle ✓

= 2 × Area of kite

$= 2 \times 3 = 6 \text{ m}^2$

so DB × AC = 6

so DB × 2·5 = 6 ✓

$\Rightarrow \text{DB} = \frac{6}{2{\cdot}5} = \frac{12}{5} = 2{\cdot}4 \text{ m}$ ✓

3 marks

Strategy

- For this mark you will need to show you understood the relationship between the diagonals of the kite and the area of the kite.

Equation

- Product of the two diagonals = 2 × Area of kite leads to DB × 2·5 = 6 which gains this mark.

Solution

- Divide both sides by 2·5.
- Doubling top and bottom gets rid of the problem of division by 2·5.

NOTES: 2·1 page 13

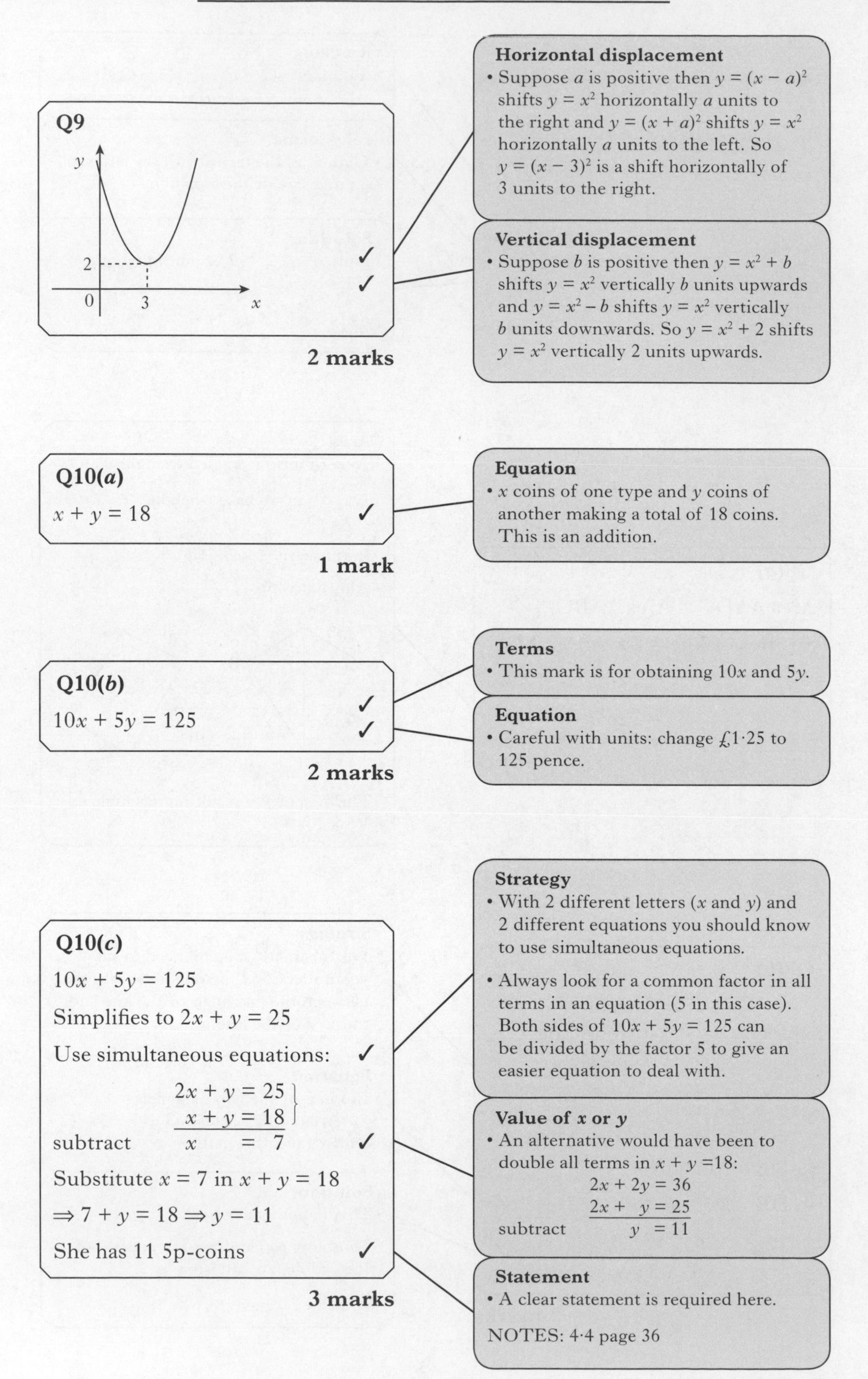

Q9
y
2
0
3
x
✓
✓
2 marks
Horizontal displacement
• Suppose a is positive then $y = (x - a)^2$ shifts $y = x^2$ horizontally a units to the right and $y = (x + a)^2$ shifts $y = x^2$ horizontally a units to the left. So $y = (x - 3)^2$ is a shift horizontally of 3 units to the right.
Vertical displacement
• Suppose b is positive then $y = x^2 + b$ shifts $y = x^2$ vertically b units upwards and $y = x^2 - b$ shifts $y = x^2$ vertically b units downwards. So $y = x^2 + 2$ shifts $y = x^2$ vertically 2 units upwards.
Q10(a)
$x + y = 18$ ✓
1 mark
Equation
• x coins of one type and y coins of another making a total of 18 coins. This is an addition.
Q10(b)
$10x + 5y = 125$ ✓ ✓
2 marks
Terms
• This mark is for obtaining $10x$ and $5y$.
Equation
• Careful with units: change £1·25 to 125 pence.
Q10(c)
$10x + 5y = 125$
Simplifies to $2x + y = 25$
Use simultaneous equations: ✓
$2x + y = 25$
$x + y = 18$
subtract $x = 7$ ✓
Substitute $x = 7$ in $x + y = 18$
$\Rightarrow 7 + y = 18 \Rightarrow y = 11$
She has 11 5p-coins ✓
3 marks
Strategy
• With 2 different letters (x and y) and 2 different equations you should know to use simultaneous equations.
• Always look for a common factor in all terms in an equation (5 in this case). Both sides of $10x + 5y = 125$ can be divided by the factor 5 to give an easier equation to deal with.
Value of x or y
• An alternative would have been to double all terms in $x + y = 18$:
$2x + 2y = 36$
$2x + y = 25$
subtract $y = 11$
Statement
• A clear statement is required here.
NOTES: 4·4 page 36

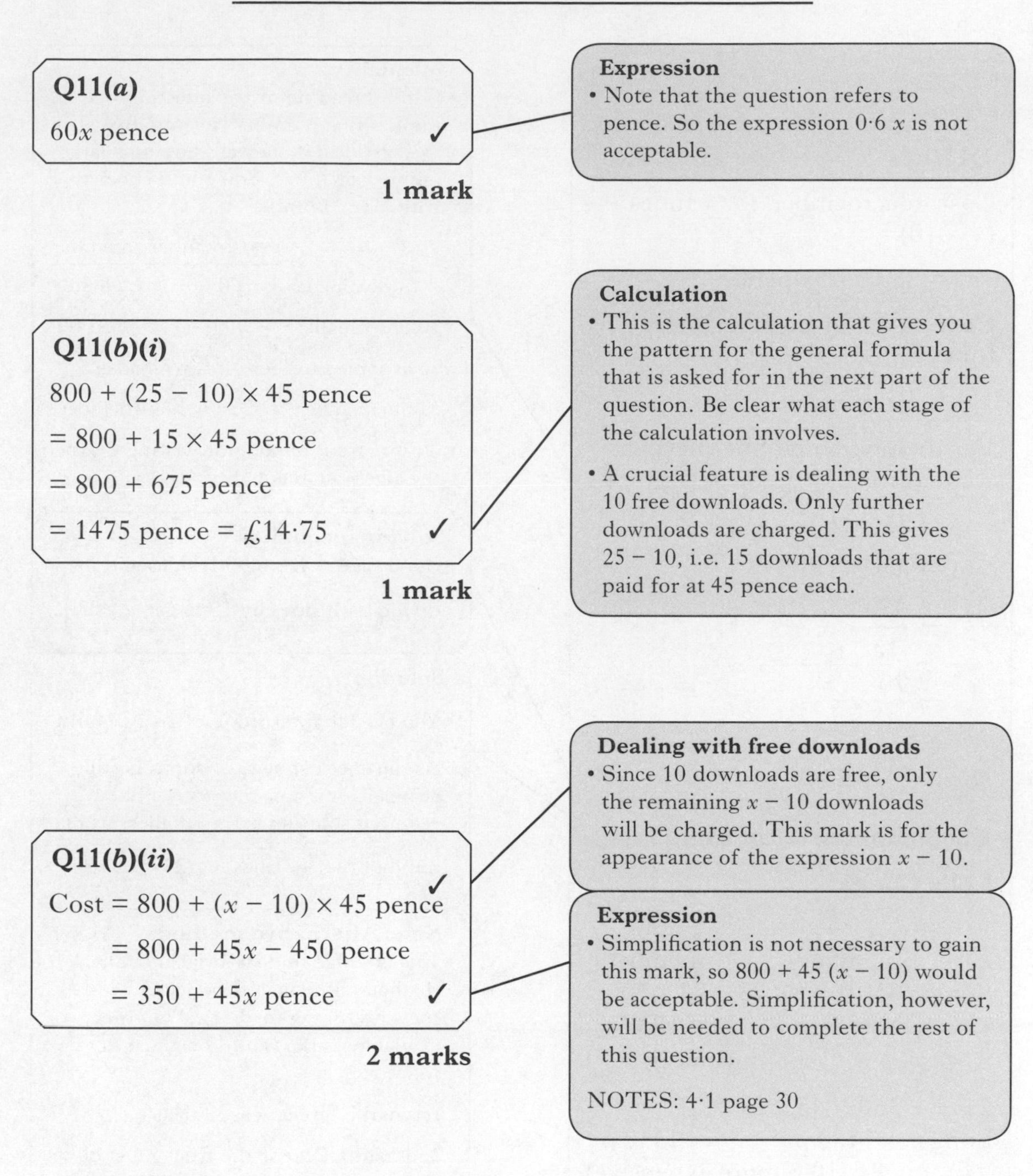

Q11(a)

$60x$ pence ✓

1 mark

Expression
- Note that the question refers to pence. So the expression $0{\cdot}6\,x$ is not acceptable.

Q11(b)(i)

$800 + (25 - 10) \times 45$ pence

$= 800 + 15 \times 45$ pence

$= 800 + 675$ pence

$= 1475$ pence $= £14{\cdot}75$ ✓

1 mark

Calculation
- This is the calculation that gives you the pattern for the general formula that is asked for in the next part of the question. Be clear what each stage of the calculation involves.
- A crucial feature is dealing with the 10 free downloads. Only further downloads are charged. This gives 25 − 10, i.e. 15 downloads that are paid for at 45 pence each.

Q11(b)(ii)

Cost $= 800 + (x - 10) \times 45$ pence ✓

$= 800 + 45x - 450$ pence

$= 350 + 45x$ pence ✓

2 marks

Dealing with free downloads
- Since 10 downloads are free, only the remaining $x - 10$ downloads will be charged. This mark is for the appearance of the expression $x - 10$.

Expression
- Simplification is not necessary to gain this mark, so $800 + 45\,(x - 10)$ would be acceptable. Simplification, however, will be needed to complete the rest of this question.

NOTES: 4·1 page 30

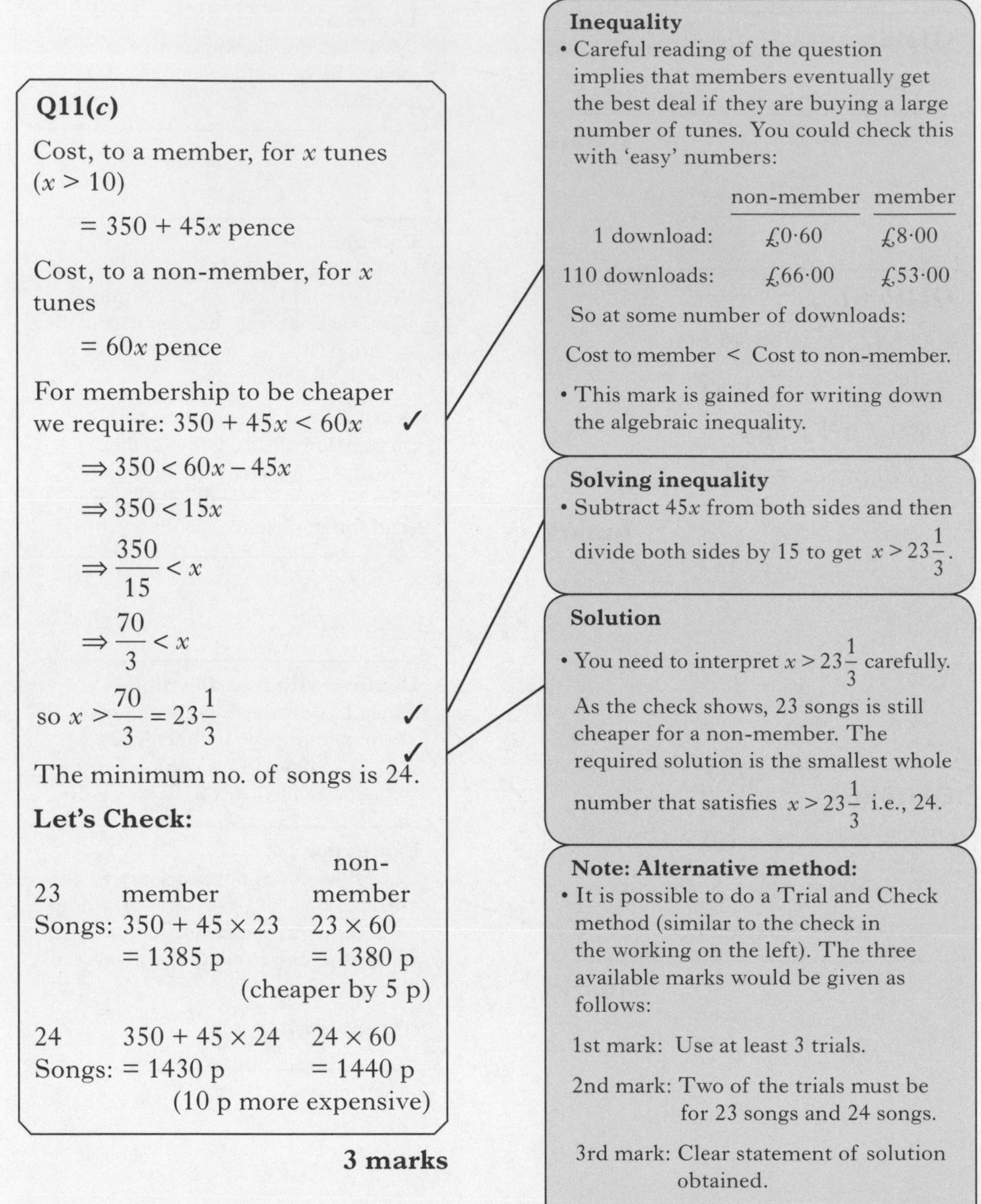

Q11(*c*)

Cost, to a member, for x tunes ($x > 10$)

$= 350 + 45x$ pence

Cost, to a non-member, for x tunes

$= 60x$ pence

For membership to be cheaper we require: $350 + 45x < 60x$ ✓

$\Rightarrow 350 < 60x - 45x$

$\Rightarrow 350 < 15x$

$\Rightarrow \frac{350}{15} < x$

$\Rightarrow \frac{70}{3} < x$

so $x > \frac{70}{3} = 23\frac{1}{3}$ ✓

The minimum no. of songs is 24. ✓

Let's Check:

	member	non-member
23 Songs:	$350 + 45 \times 23$ = 1385 p	23×60 = 1380 p (cheaper by 5 p)
24 Songs:	$350 + 45 \times 24$ = 1430 p	24×60 = 1440 p (10 p more expensive)

3 marks

Inequality

- Careful reading of the question implies that members eventually get the best deal if they are buying a large number of tunes. You could check this with 'easy' numbers:

	non-member	member
1 download:	£0·60	£8·00
110 downloads:	£66·00	£53·00

So at some number of downloads:

Cost to member < Cost to non-member.

- This mark is gained for writing down the algebraic inequality.

Solving inequality

- Subtract $45x$ from both sides and then divide both sides by 15 to get $x > 23\frac{1}{3}$.

Solution

- You need to interpret $x > 23\frac{1}{3}$ carefully.

As the check shows, 23 songs is still cheaper for a non-member. The required solution is the smallest whole number that satisfies $x > 23\frac{1}{3}$ i.e., 24.

Note: Alternative method:

- It is possible to do a Trial and Check method (similar to the check in the working on the left). The three available marks would be given as follows:

1st mark: Use at least 3 trials.

2nd mark: Two of the trials must be for 23 songs and 24 songs.

3rd mark: Clear statement of solution obtained.

NOTES: 4·4 page 35

Practice Exam F: Paper 2 Worked Answers

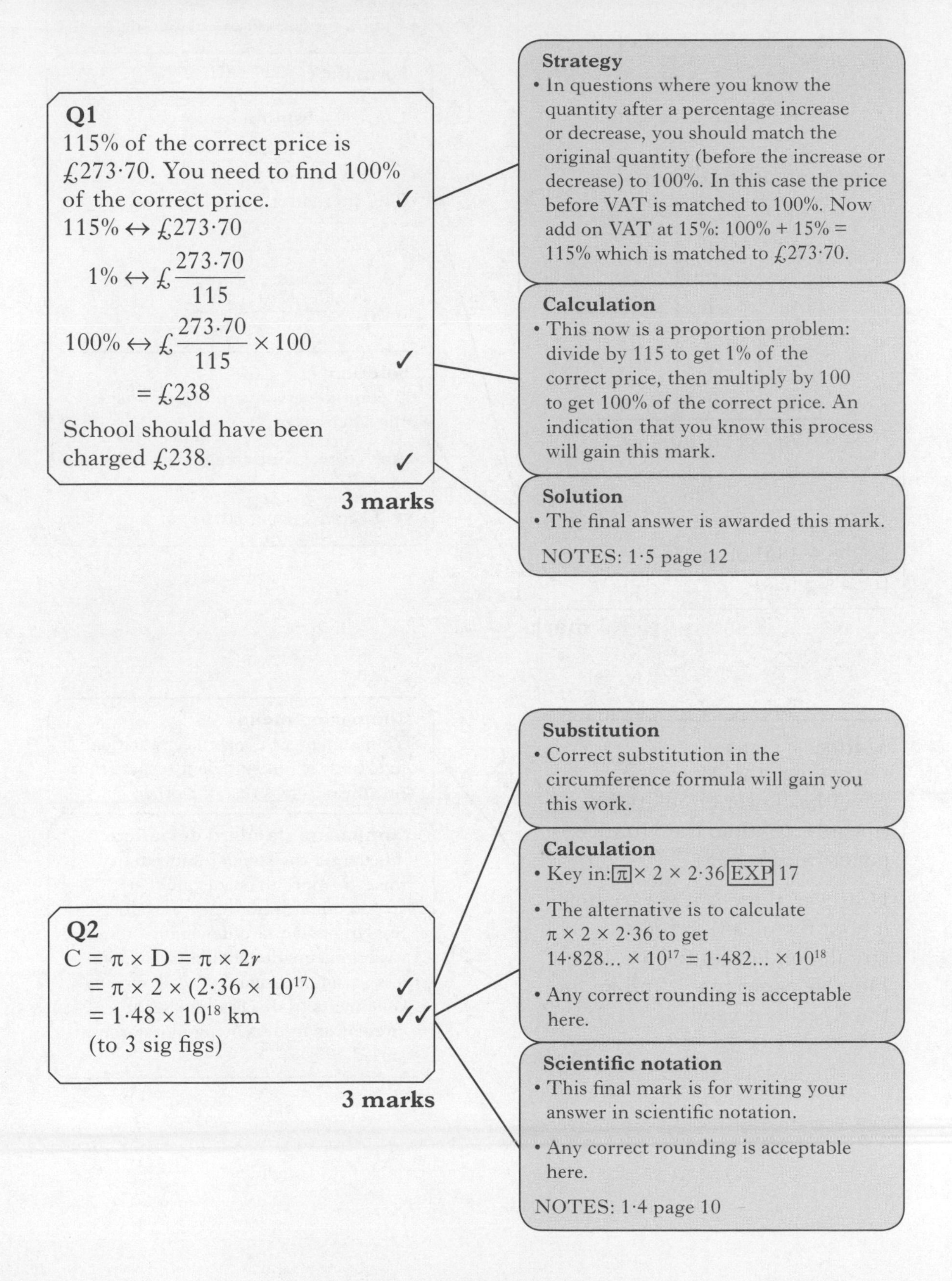

Q1

115% of the correct price is £273·70. You need to find 100% of the correct price. ✓

115% ↔ £273·70

$1\% \leftrightarrow £\frac{273 \cdot 70}{115}$

$100\% \leftrightarrow £\frac{273 \cdot 70}{115} \times 100$ ✓

$= £238$

School should have been charged £238. ✓

3 marks

Strategy
- In questions where you know the quantity after a percentage increase or decrease, you should match the original quantity (before the increase or decrease) to 100%. In this case the price before VAT is matched to 100%. Now add on VAT at 15%: 100% + 15% = 115% which is matched to £273·70.

Calculation
- This now is a proportion problem: divide by 115 to get 1% of the correct price, then multiply by 100 to get 100% of the correct price. An indication that you know this process will gain this mark.

Solution
- The final answer is awarded this mark.

NOTES: 1·5 page 12

Q2

$C = \pi \times D = \pi \times 2r$

$= \pi \times 2 \times (2 \cdot 36 \times 10^{17})$ ✓

$= 1 \cdot 48 \times 10^{18}$ km ✓✓

(to 3 sig figs)

3 marks

Substitution
- Correct substitution in the circumference formula will gain you this work.

Calculation
- Key in: [π] × 2 × 2·36 [EXP] 17
- The alternative is to calculate $\pi \times 2 \times 2 \cdot 36$ to get $14 \cdot 828... \times 10^{17} = 1 \cdot 482... \times 10^{18}$
- Any correct rounding is acceptable here.

Scientific notation
- This final mark is for writing your answer in scientific notation.
- Any correct rounding is acceptable here.

NOTES: 1·4 page 10

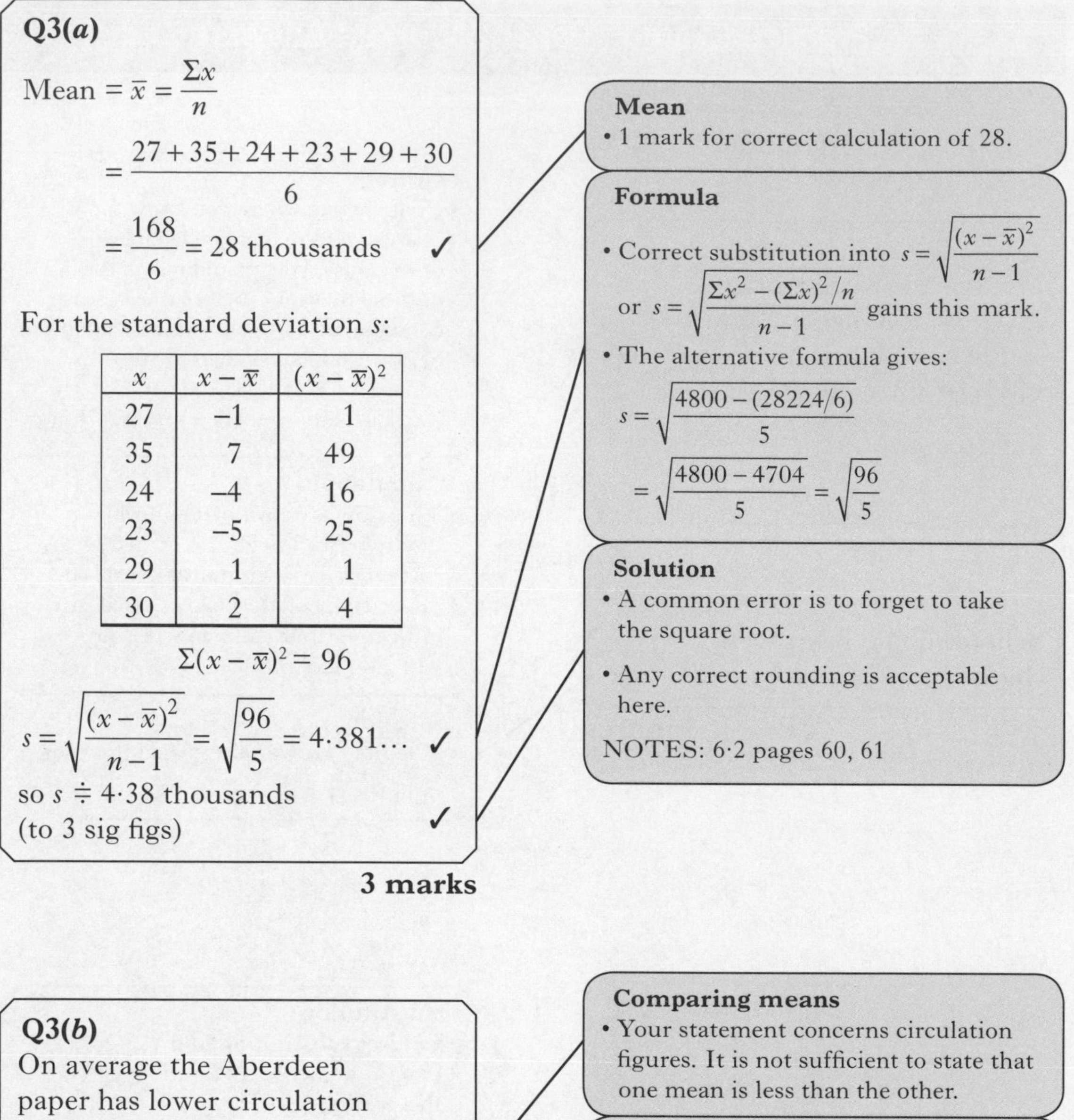

Q3(*a*)

Mean $= \overline{x} = \dfrac{\Sigma x}{n}$

$= \dfrac{27+35+24+23+29+30}{6}$

$= \dfrac{168}{6} = 28$ thousands ✓

For the standard deviation s:

x	$x - \overline{x}$	$(x - \overline{x})^2$
27	–1	1
35	7	49
24	–4	16
23	–5	25
29	1	1
30	2	4

$\Sigma(x - \overline{x})^2 = 96$

$s = \sqrt{\dfrac{(x-\overline{x})^2}{n-1}} = \sqrt{\dfrac{96}{5}} = 4{\cdot}381\ldots$ ✓

so $s \doteqdot 4{\cdot}38$ thousands (to 3 sig figs) ✓

3 marks

Mean
- 1 mark for correct calculation of 28.

Formula
- Correct substitution into $s = \sqrt{\dfrac{(x-\overline{x})^2}{n-1}}$ or $s = \sqrt{\dfrac{\Sigma x^2 - (\Sigma x)^2/n}{n-1}}$ gains this mark.
- The alternative formula gives:

$s = \sqrt{\dfrac{4800 - (28224/6)}{5}}$

$= \sqrt{\dfrac{4800 - 4704}{5}} = \sqrt{\dfrac{96}{5}}$

Solution
- A common error is to forget to take the square root.
- Any correct rounding is acceptable here.

NOTES: 6·2 pages 60, 61

Q3(*b*)

On average the Aberdeen paper has lower circulation (mean = 28) than the Dundee paper (mean = 35). ✓

However, there is less variation (about the mean) in the circulation figures for the Dundee paper ($s = 1{\cdot}2$) than for the Aberdeen paper ($s = 4{\cdot}38$) ✓

2 marks

Comparing means
- Your statement concerns circulation figures. It is not sufficient to state that one mean is less than the other.

Comparing standard deviations
- The larger the standard deviation then the more the data values are spread out around the mean value. So: larger standard deviation – more variation, smaller standard deviation – less variation. Again, you are making statements in this case about the circulation figures being more or less varied.

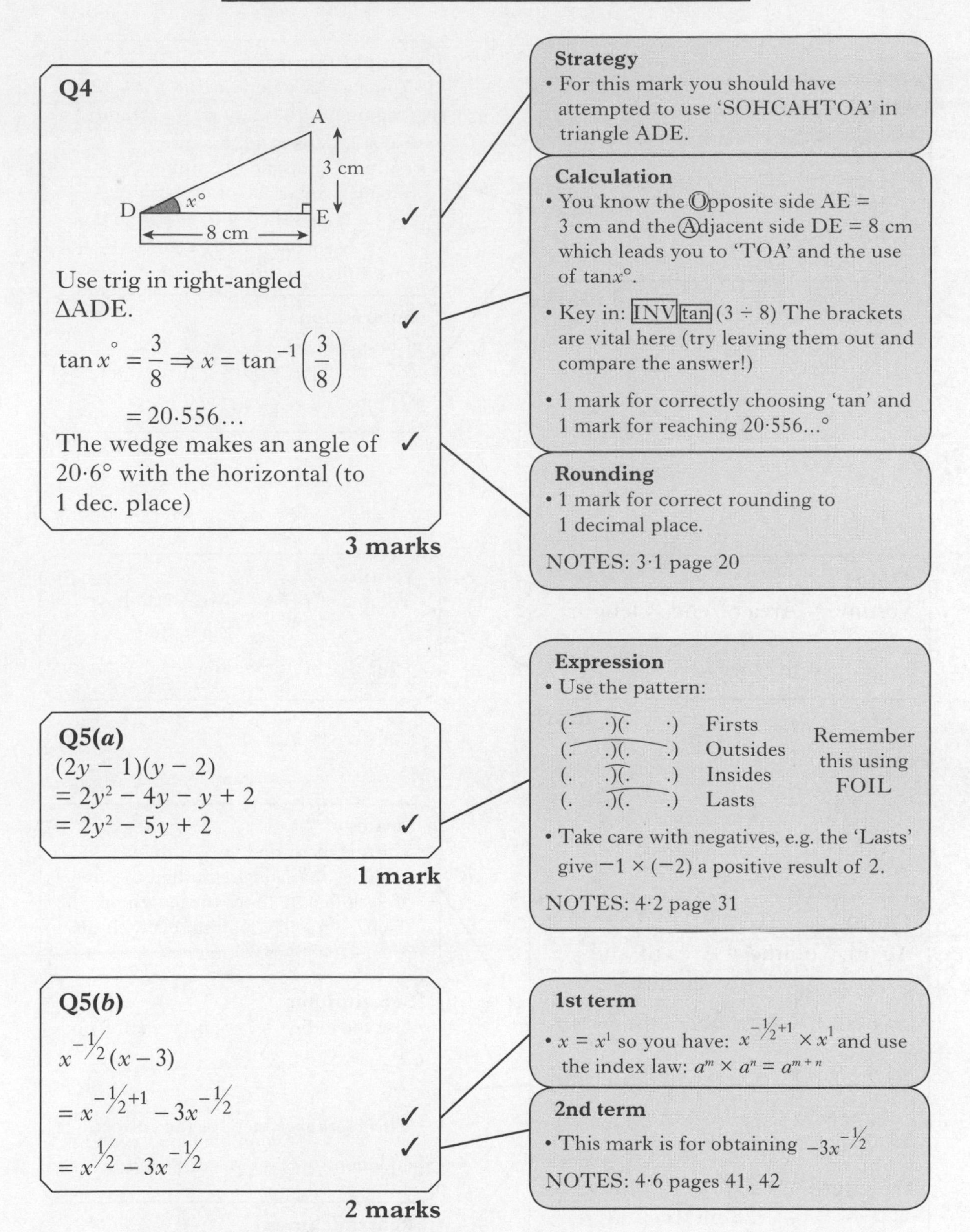

Q4

Use trig in right-angled ΔADE. ✓

$$\tan x° = \frac{3}{8} \Rightarrow x = \tan^{-1}\left(\frac{3}{8}\right)$$

$$= 20{\cdot}556\ldots$$ ✓

The wedge makes an angle of $20{\cdot}6°$ with the horizontal (to 1 dec. place) ✓

3 marks

Strategy

- For this mark you should have attempted to use 'SOHCAHTOA' in triangle ADE.

Calculation

- You know the Ⓞpposite side AE = 3 cm and the Ⓐdjacent side DE = 8 cm which leads you to 'TOA' and the use of $\tan x°$.
- Key in: INV tan $(3 \div 8)$ The brackets are vital here (try leaving them out and compare the answer!)
- 1 mark for correctly choosing 'tan' and 1 mark for reaching $20{\cdot}556\ldots°$

Rounding

- 1 mark for correct rounding to 1 decimal place.

NOTES: 3·1 page 20

Q5(*a*)

$(2y - 1)(y - 2)$
$= 2y^2 - 4y - y + 2$
$= 2y^2 - 5y + 2$ ✓

1 mark

Expression

- Use the pattern:

(·)(·)	Firsts	Remember this using FOIL
(.)(.)	Outsides	
(.)(.)	Insides	
(.)(.)	Lasts	

- Take care with negatives, e.g. the 'Lasts' give $-1 \times (-2)$ a positive result of 2.

NOTES: 4·2 page 31

Q5(*b*)

$x^{-\frac{1}{2}}(x - 3)$

$= x^{-\frac{1}{2}+1} - 3x^{-\frac{1}{2}}$ ✓

$= x^{\frac{1}{2}} - 3x^{-\frac{1}{2}}$ ✓

2 marks

1st term

- $x = x^1$ so you have: $x^{-\frac{1}{2}+1} \times x^1$ and use the index law: $a^m \times a^n = a^{m+n}$

2nd term

- This mark is for obtaining $-3x^{-\frac{1}{2}}$

NOTES: 4·6 pages 41, 42

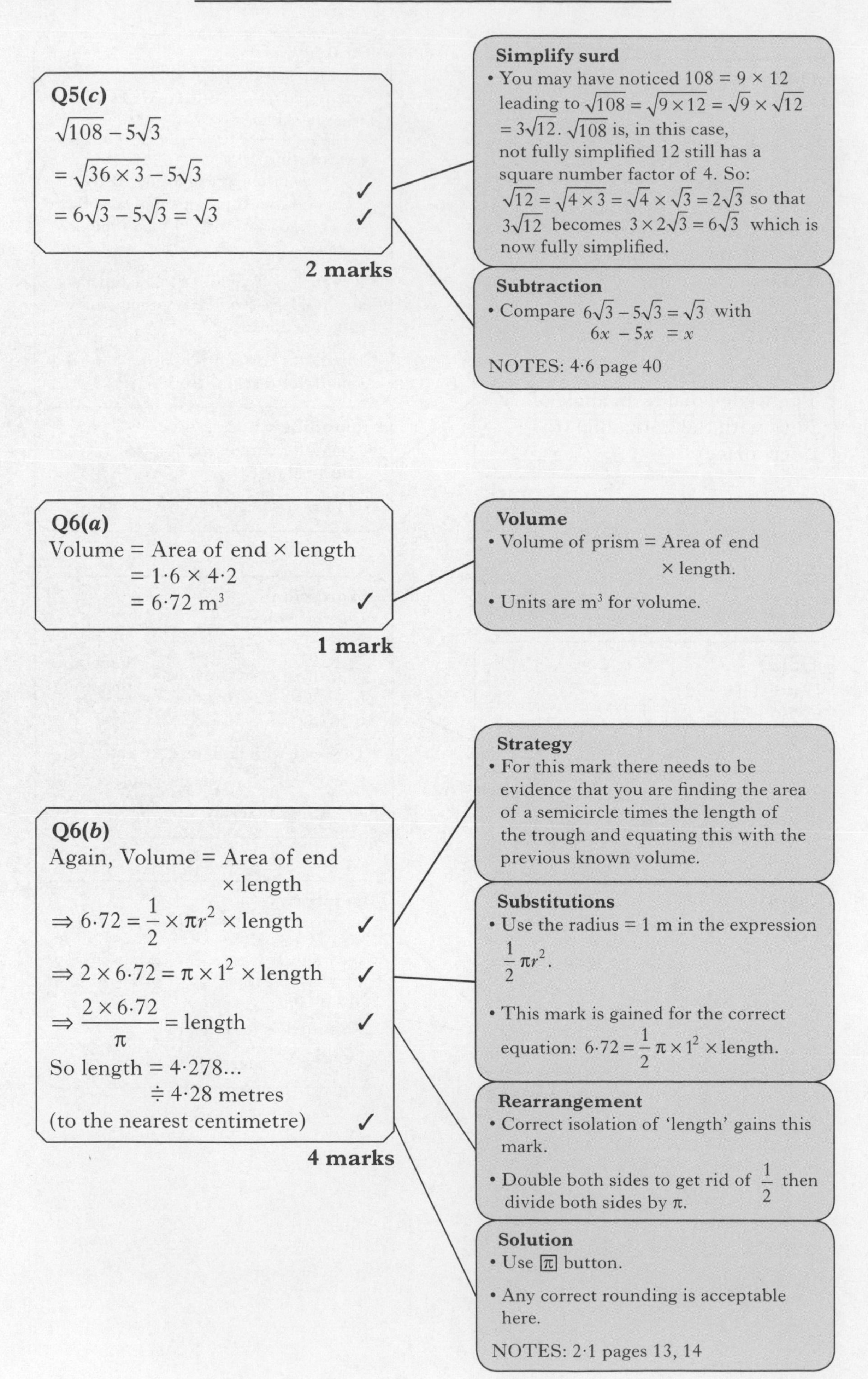

Q5(*c*)

$\sqrt{108} - 5\sqrt{3}$

$= \sqrt{36 \times 3} - 5\sqrt{3}$ ✓

$= 6\sqrt{3} - 5\sqrt{3} = \sqrt{3}$ ✓

2 marks

Simplify surd

- You may have noticed $108 = 9 \times 12$ leading to $\sqrt{108} = \sqrt{9 \times 12} = \sqrt{9} \times \sqrt{12} = 3\sqrt{12}$. $\sqrt{108}$ is, in this case, not fully simplified 12 still has a square number factor of 4. So: $\sqrt{12} = \sqrt{4 \times 3} = \sqrt{4} \times \sqrt{3} = 2\sqrt{3}$ so that $3\sqrt{12}$ becomes $3 \times 2\sqrt{3} = 6\sqrt{3}$ which is now fully simplified.

Subtraction

- Compare $6\sqrt{3} - 5\sqrt{3} = \sqrt{3}$ with $6x - 5x = x$

NOTES: 4·6 page 40

Q6(*a*)

Volume = Area of end × length

$= 1{\cdot}6 \times 4{\cdot}2$

$= 6{\cdot}72 \text{ m}^3$ ✓

1 mark

Volume

- Volume of prism = Area of end × length.
- Units are m^3 for volume.

Q6(*b*)

Again, Volume = Area of end × length

$\Rightarrow 6{\cdot}72 = \frac{1}{2} \times \pi r^2 \times \text{length}$ ✓

$\Rightarrow 2 \times 6{\cdot}72 = \pi \times 1^2 \times \text{length}$ ✓

$\Rightarrow \frac{2 \times 6{\cdot}72}{\pi} = \text{length}$ ✓

So length = 4·278...

≑ 4·28 metres

(to the nearest centimetre) ✓

4 marks

Strategy

- For this mark there needs to be evidence that you are finding the area of a semicircle times the length of the trough and equating this with the previous known volume.

Substitutions

- Use the radius = 1 m in the expression $\frac{1}{2}\pi r^2$.
- This mark is gained for the correct equation: $6{\cdot}72 = \frac{1}{2}\pi \times 1^2 \times \text{length}$.

Rearrangement

- Correct isolation of 'length' gains this mark.
- Double both sides to get rid of $\frac{1}{2}$ then divide both sides by π.

Solution

- Use [π] button.
- Any correct rounding is acceptable here.

NOTES: 2·1 pages 13, 14

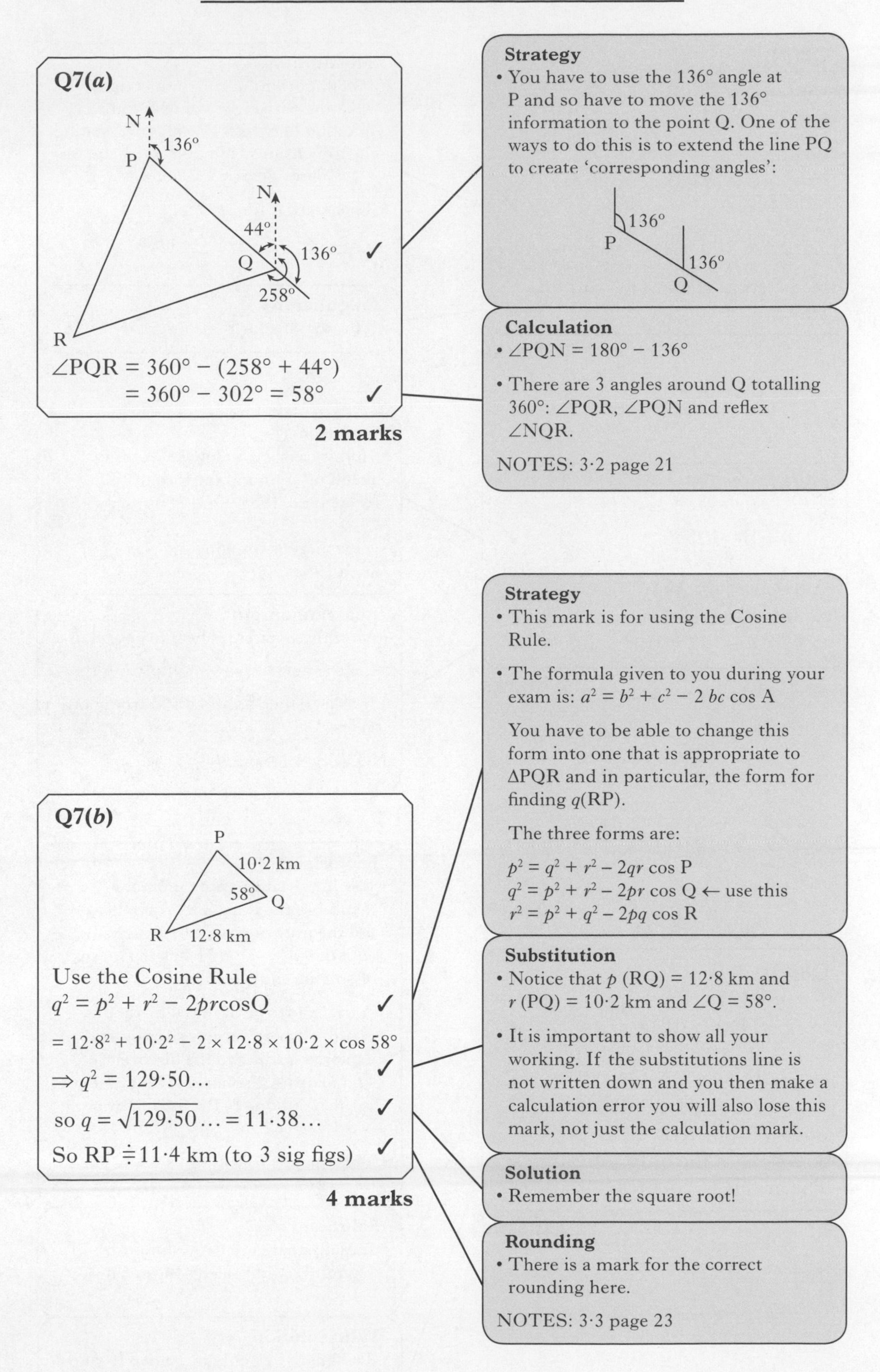

Q7(*a*)

$\angle PQR = 360° - (258° + 44°)$ ✓

$= 360° - 302° = 58°$ ✓

2 marks

Strategy
- You have to use the 136° angle at P and so have to move the 136° information to the point Q. One of the ways to do this is to extend the line PQ to create 'corresponding angles':

Calculation
- $\angle PQN = 180° - 136°$
- There are 3 angles around Q totalling 360°: $\angle PQR$, $\angle PQN$ and reflex $\angle NQR$.

NOTES: 3·2 page 21

Q7(*b*)

Use the Cosine Rule

$q^2 = p^2 + r^2 - 2pr\cos Q$ ✓

$= 12{\cdot}8^2 + 10{\cdot}2^2 - 2 \times 12{\cdot}8 \times 10{\cdot}2 \times \cos 58°$ ✓

$\Rightarrow q^2 = 129{\cdot}50...$ ✓

so $q = \sqrt{129{\cdot}50...} = 11{\cdot}38...$ ✓

So RP $\doteqdot$ 11·4 km (to 3 sig figs) ✓

4 marks

Strategy
- This mark is for using the Cosine Rule.
- The formula given to you during your exam is: $a^2 = b^2 + c^2 - 2\,bc\cos A$

You have to be able to change this form into one that is appropriate to ΔPQR and in particular, the form for finding q(RP).

The three forms are:

$p^2 = q^2 + r^2 - 2qr\cos P$
$q^2 = p^2 + r^2 - 2pr\cos Q$ ← use this
$r^2 = p^2 + q^2 - 2pq\cos R$

Substitution
- Notice that p (RQ) = 12·8 km and r (PQ) = 10·2 km and $\angle Q = 58°$.
- It is important to show all your working. If the substitutions line is not written down and you then make a calculation error you will also lose this mark, not just the calculation mark.

Solution
- Remember the square root!

Rounding
- There is a mark for the correct rounding here.

NOTES: 3·3 page 23

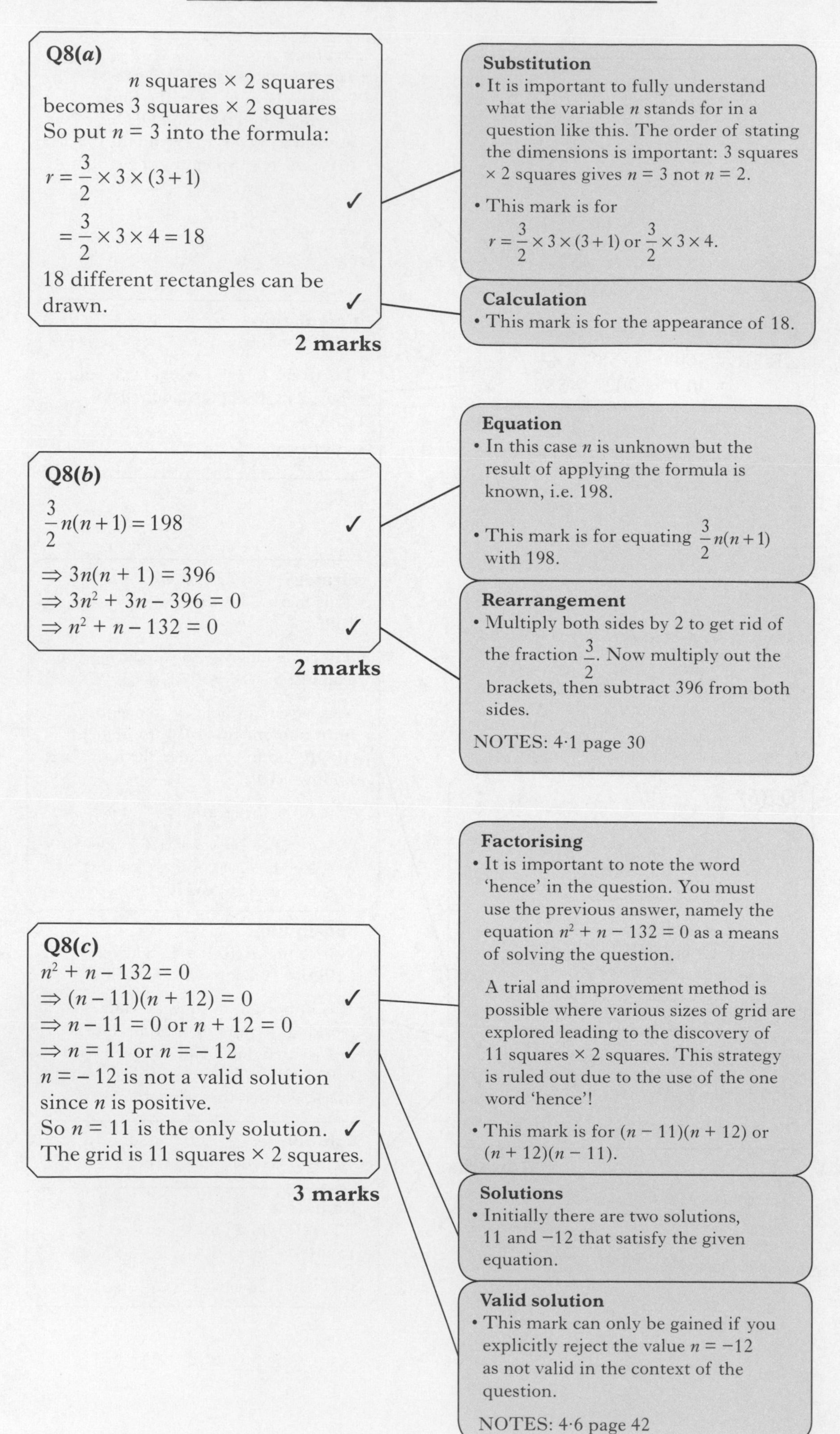

Q8(*a*)

n squares × 2 squares
becomes 3 squares × 2 squares
So put $n = 3$ into the formula:

$r = \frac{3}{2} \times 3 \times (3+1)$ ✓

$= \frac{3}{2} \times 3 \times 4 = 18$

18 different rectangles can be drawn. ✓

2 marks

Substitution
- It is important to fully understand what the variable n stands for in a question like this. The order of stating the dimensions is important: 3 squares × 2 squares gives $n = 3$ not $n = 2$.
- This mark is for $r = \frac{3}{2} \times 3 \times (3+1)$ or $\frac{3}{2} \times 3 \times 4$.

Calculation
- This mark is for the appearance of 18.

Q8(*b*)

$\frac{3}{2}n(n+1) = 198$ ✓

$\Rightarrow 3n(n + 1) = 396$

$\Rightarrow 3n^2 + 3n - 396 = 0$

$\Rightarrow n^2 + n - 132 = 0$ ✓

2 marks

Equation
- In this case n is unknown but the result of applying the formula is known, i.e. 198.
- This mark is for equating $\frac{3}{2}n(n+1)$ with 198.

Rearrangement
- Multiply both sides by 2 to get rid of the fraction $\frac{3}{2}$. Now multiply out the brackets, then subtract 396 from both sides.

NOTES: 4·1 page 30

Q8(*c*)

$n^2 + n - 132 = 0$

$\Rightarrow (n - 11)(n + 12) = 0$ ✓

$\Rightarrow n - 11 = 0$ or $n + 12 = 0$

$\Rightarrow n = 11$ or $n = -12$ ✓

$n = -12$ is not a valid solution since n is positive.

So $n = 11$ is the only solution. ✓

The grid is 11 squares × 2 squares.

3 marks

Factorising
- It is important to note the word 'hence' in the question. You must use the previous answer, namely the equation $n^2 + n - 132 = 0$ as a means of solving the question.

  A trial and improvement method is possible where various sizes of grid are explored leading to the discovery of 11 squares × 2 squares. This strategy is ruled out due to the use of the one word 'hence'!
- This mark is for $(n - 11)(n + 12)$ or $(n + 12)(n - 11)$.

Solutions
- Initially there are two solutions, 11 and −12 that satisfy the given equation.

Valid solution
- This mark can only be gained if you explicitly reject the value $n = -12$ as not valid in the context of the question.

NOTES: 4·6 page 42

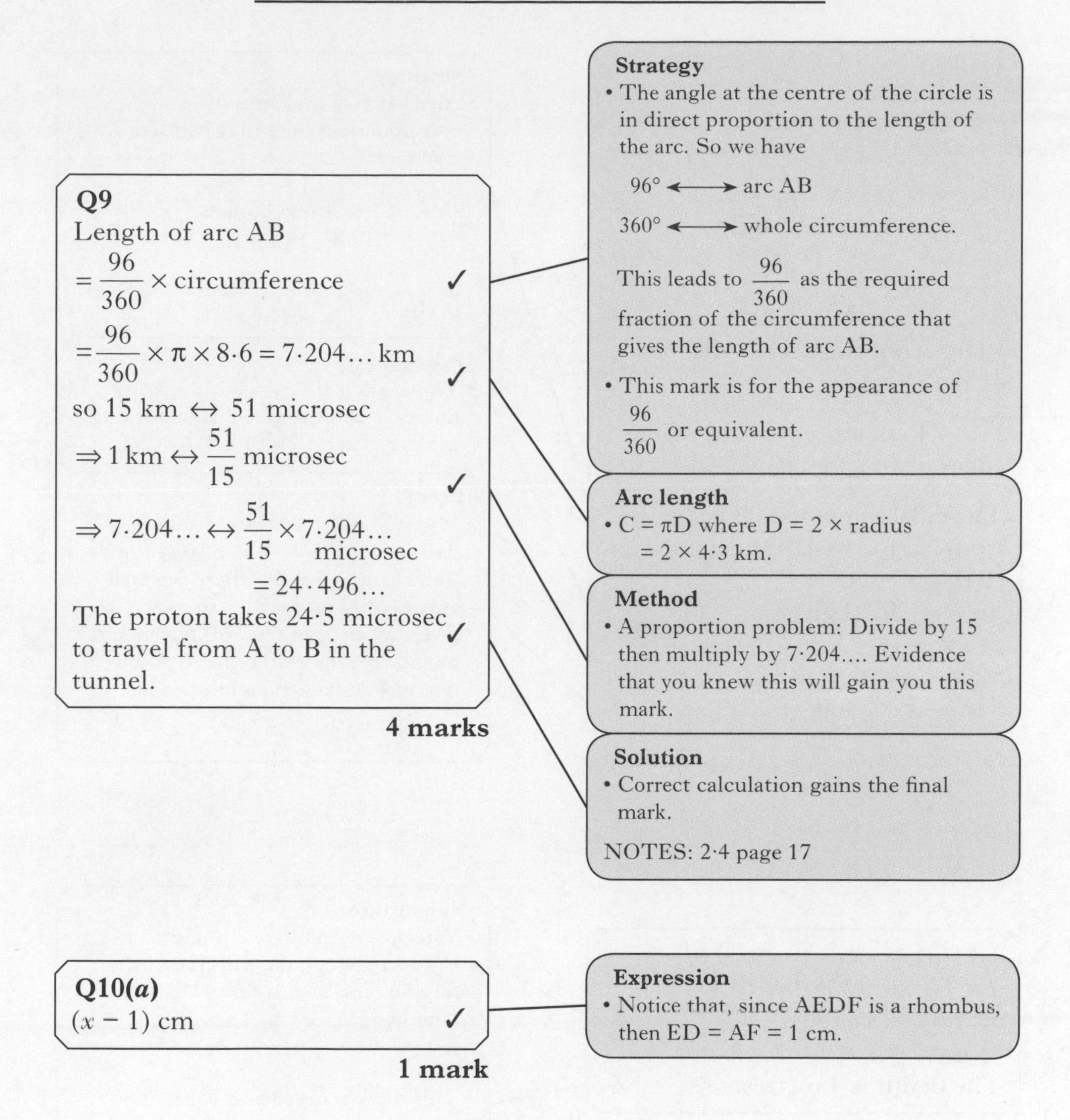

Q9

Length of arc AB

$= \frac{96}{360} \times \text{circumference}$ ✓

$= \frac{96}{360} \times \pi \times 8{\cdot}6 = 7{\cdot}204\ldots \text{ km}$ ✓

so 15 km ↔ 51 microsec

$\Rightarrow 1 \text{ km} \leftrightarrow \frac{51}{15}$ microsec ✓

$\Rightarrow 7{\cdot}204\ldots \leftrightarrow \frac{51}{15} \times 7{\cdot}204\ldots$ microsec

$= 24{\cdot}496\ldots$

The proton takes 24·5 microsec to travel from A to B in the tunnel. ✓

4 marks

Strategy

- The angle at the centre of the circle is in direct proportion to the length of the arc. So we have

  96° ⟷ arc AB

  360° ⟷ whole circumference.

  This leads to $\frac{96}{360}$ as the required fraction of the circumference that gives the length of arc AB.
- This mark is for the appearance of $\frac{96}{360}$ or equivalent.

Arc length

- $C = \pi D$ where $D = 2 \times \text{radius}$ $= 2 \times 4{\cdot}3$ km.

Method

- A proportion problem: Divide by 15 then multiply by 7·204.... Evidence that you knew this will gain you this mark.

Solution

- Correct calculation gains the final mark.

NOTES: 2·4 page 17

Q10(*a*)

$(x - 1)$ cm ✓

1 mark

Expression

- Notice that, since AEDF is a rhombus, then ED = AF = 1 cm.

Q10(*b*)

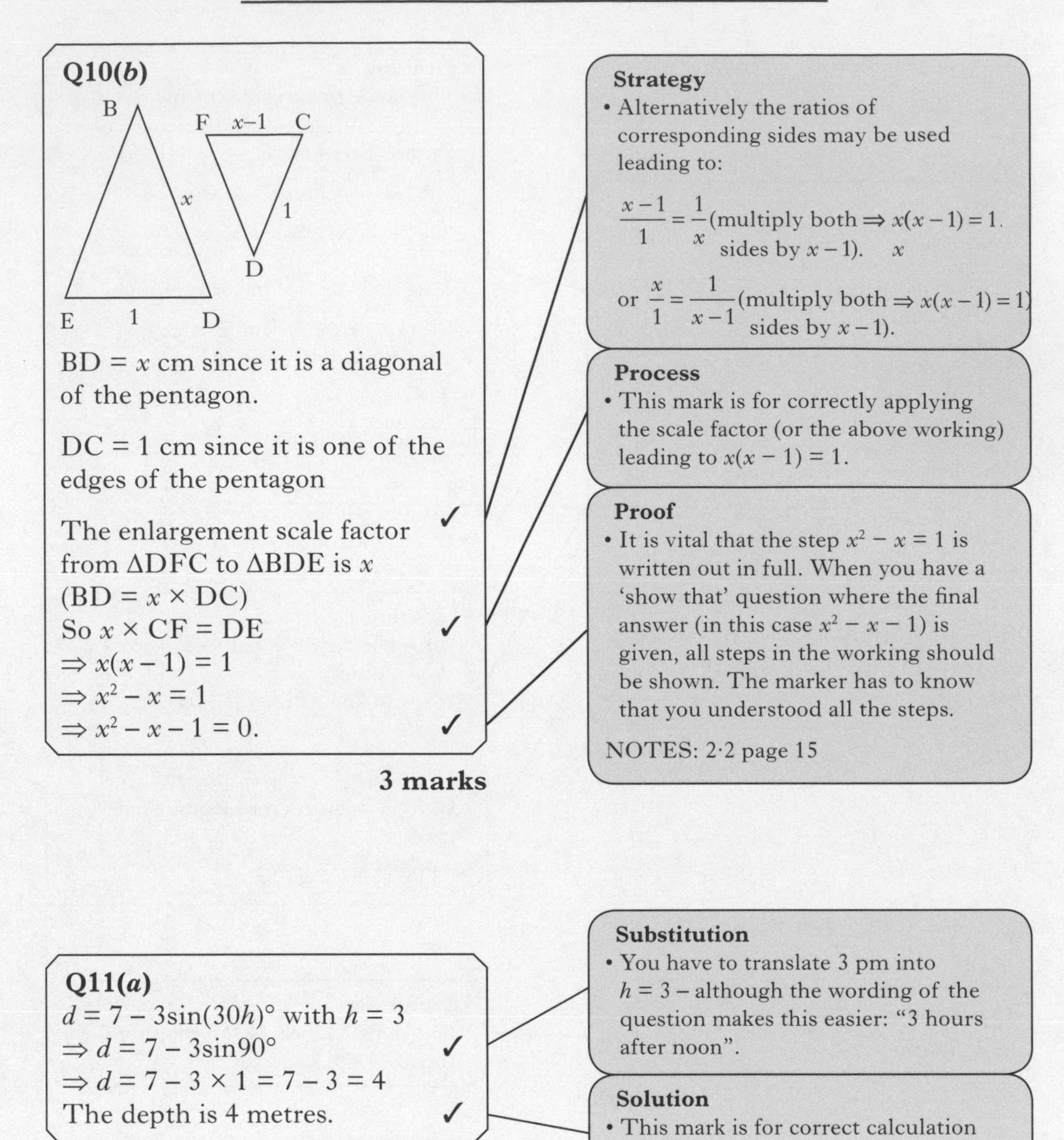

BD = x cm since it is a diagonal of the pentagon.

DC = 1 cm since it is one of the edges of the pentagon

The enlargement scale factor from ΔDFC to ΔBDE is x ✓
(BD = $x \times$ DC)
So $x \times$ CF = DE ✓
$\Rightarrow x(x-1) = 1$
$\Rightarrow x^2 - x = 1$
$\Rightarrow x^2 - x - 1 = 0.$ ✓

3 marks

Strategy

- Alternatively the ratios of corresponding sides may be used leading to:

$\frac{x-1}{1} = \frac{1}{x}$ (multiply both sides by $x - 1$). $\Rightarrow x(x-1) = 1.$ x

or $\frac{x}{1} = \frac{1}{x-1}$ (multiply both sides by $x - 1$). $\Rightarrow x(x-1) = 1$

Process

- This mark is for correctly applying the scale factor (or the above working) leading to $x(x-1) = 1$.

Proof

- It is vital that the step $x^2 - x = 1$ is written out in full. When you have a 'show that' question where the final answer (in this case $x^2 - x - 1$) is given, all steps in the working should be shown. The marker has to know that you understood all the steps.

NOTES: 2·2 page 15

Q11(*a*)

$d = 7 - 3\sin(30h)°$ with $h = 3$
$\Rightarrow d = 7 - 3\sin 90°$ ✓
$\Rightarrow d = 7 - 3 \times 1 = 7 - 3 = 4$
The depth is 4 metres. ✓

2 marks

Substitution

- You have to translate 3 pm into $h = 3$ – although the wording of the question makes this easier: "3 hours after noon".

Solution

- This mark is for correct calculation of 4.

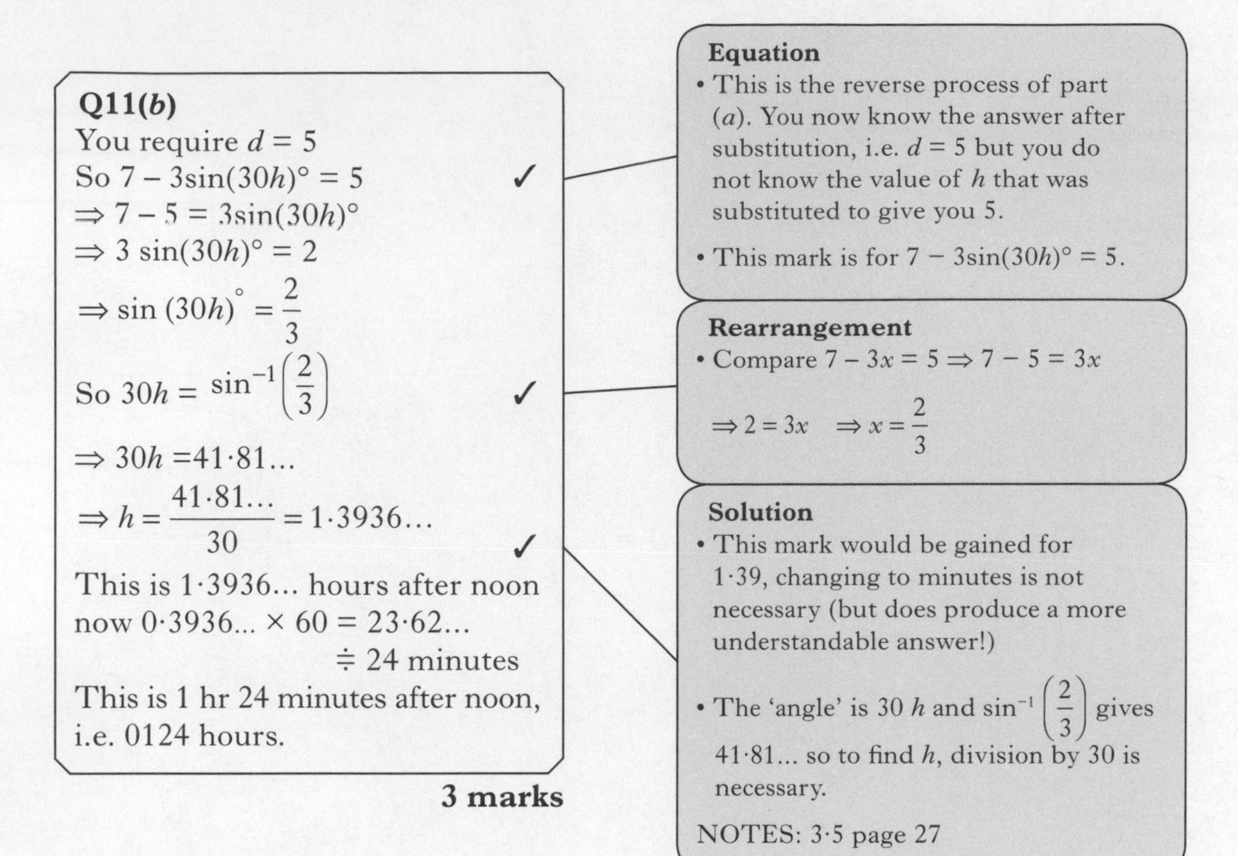

Q11(*b*)

You require $d = 5$

So $7 - 3\sin(30h)^\circ = 5$ ✓

$\Rightarrow 7 - 5 = 3\sin(30h)^\circ$

$\Rightarrow 3\sin(30h)^\circ = 2$

$\Rightarrow \sin(30h)^\circ = \frac{2}{3}$

So $30h = \sin^{-1}\left(\frac{2}{3}\right)$ ✓

$\Rightarrow 30h = 41{\cdot}81\ldots$

$\Rightarrow h = \frac{41{\cdot}81\ldots}{30} = 1{\cdot}3936\ldots$ ✓

This is 1·3936... hours after noon

now $0{\cdot}3936\ldots \times 60 = 23{\cdot}62\ldots$

$\doteqdot 24$ minutes

This is 1 hr 24 minutes after noon, i.e. 0124 hours.

3 marks

Equation

- This is the reverse process of part (*a*). You now know the answer after substitution, i.e. $d = 5$ but you do not know the value of h that was substituted to give you 5.
- This mark is for $7 - 3\sin(30h)^\circ = 5$.

Rearrangement

- Compare $7 - 3x = 5 \Rightarrow 7 - 5 = 3x$

  $\Rightarrow 2 = 3x \quad \Rightarrow x = \frac{2}{3}$

Solution

- This mark would be gained for 1·39, changing to minutes is not necessary (but does produce a more understandable answer!)
- The 'angle' is 30 h and $\sin^{-1}\left(\frac{2}{3}\right)$ gives 41·81... so to find h, division by 30 is necessary.

NOTES: 3·5 page 27